GREEN ENERGY

GREEN ENERGY

Sustainable Electricity Supply with Low Environmental Impact

Eric Jeffs

CRC Press
Taylor & Francis Group
Boca Raton London New York

CRC Press is an imprint of the
Taylor & Francis Group, an **informa** business

CRC Press
Taylor & Francis Group
6000 Broken Sound Parkway NW, Suite 300
Boca Raton, FL 33487-2742

First issued in paperback 2017

© 2010 by Taylor and Francis Group, LLC
CRC Press is an imprint of Taylor & Francis Group, an Informa business

No claim to original U.S. Government works

ISBN-13: 978-1-4398-1892-3 (hbk)
ISBN-13: 978-1-138-11367-1 (pbk)

Library of Congress Cataloging-in-Publication Data

Jeffs, Eric J.
  Green energy : sustainable electricity supply with low environmental impact / Eric Jeffs.
      p. cm.
  Includes index.
  ISBN 978-1-4398-1892-3 (hardcover : alk. paper)
    1. Electric power distribution--Environmental aspects. 2. Interconnected
  electric utility systems. 3. Renewable energy sources. 4. Sustainable engineering. I.
  Title.

  TK3001.J44 2010
  333.79'4--dc22

                                                                    2009035203

Visit the Taylor & Francis Web site at
http://www.taylorandfrancis.com

and the CRC Press Web site at
http://www.crcpress.com

CONTENTS

Acknowledgements vii

1   Introduction 1

2   Global warming 13

3   Carbon capture and storage 41

4   The end of coal? 59

5   A nuclear energy revival 83

6   Combined cycle 131

7   New energy technologies 157

8   Renewable uncertainties 169

9   The truth about America 195

10   What is in the future? 217

    Index 225

ACKNOWLEDGEMENTS

I would like to thank the many people throughout the electricity supply industries and their plant suppliers who have helped me over the years, providing information and arranging site visits around the world. In particular I would mention three people who have reviewed my manuscript and thank them for their valuable comments.

Dick Foster Pegg was born in the United Kingdom and served an apprenticeship at Rolls-Royce at the time of their early gas turbine models, before emigrating to the United States where he worked for Bechtel Corporation on power plant design, including some of the early combined cycle projects there. Later he joined Westinghouse to work on gas turbine reseach and development in Pennsylvania.

Louis Codogno was Managing Director of the Energy Division of Cockerill Mechanical Industries until his retirement in 2005. I thank him in particular for not only arranging site visits but also introducing me to many of his clients which gave me an insight into the energy policies of those countries, notably in the Middle and Far East.

Dr Maher Elmasri, was Professor of Mechanical Engineering at Massachusetts Institute of Technology until 1987, when he left to start his own company, Thermoflow Inc. It is now the leading company supplying a powerful suite of software for the design of combined cycle, and steam plants of different types including integrated gasifier combined cycles.

1

Introduction

For all but the last 250 years mankind has depended mainly on natural sources of energy. Today that would mean the sun, wind, water and biomass (fire). Where oil and coal had been found near the surface they had also been used but only as a basic source of heat, and there were fewer than 500 million people on the Earth.

The industrial revolution was founded on coal and coincided with advances in agriculture and medicine which, in little more than a century, transformed the way of life of the people of Europe and North America.

Before 1800, ships were built of wood and propelled by sails, wind and water mills ground corn and transport on land was on horseback, or in a horse-drawn carriage. As soon as coal was recognized as a fuel to generate steam the inventions followed that made it the fuel for transport and for powering industry.

At the beginning of the twentieth century, the basic elements of contemporary infrastructure were already in place. The first telephones were in use, coal-derived town gas was piped to homes, and electricity for lighting and urban transport was starting to appear on the scene in the major cities. The railway networks had been established; Charles Parsons had built the first steam turbine, which he demonstrated as the engine of a ship. Marconi had transmitted the first radio signal across the Atlantic. The first cars were on the roads. In Germany Rudolf Diesel had developed the engine which bears his name. The first powered flight was in December 1903.

The twentieth century will surely be remembered as the time when energy transformed human effort and understanding of the world as at no other time in history. But it has also been a century of war with two World Wars and several large regional conflicts. These accelerated technical developments and carried energy demand to higher levels so that in the years that followed the entire population of the developed world could have secure access to heat, light and power, and in a growing number of new applications. All of these applications depended on there being a secure supply of electricity both to produce the equipment and to operate it.

In fact the whole period since the end of the war in 1945 has been one of considerable innovation. The first transistors were produced at the Bell Telephone Laboratories in the United States in 1948, just as the first computers were starting to appear in industry. From this beginning were developed the microcircuits which are the basis of modern computer systems and process controls. It was not until 1980 that the personal computer started to appear and computing power spread into the home.

Developments in aviation, particularly the wide-bodied aircraft with large turbofan engines encouraged tourism which saw the development of hotels and various places of entertainment around the world, which all added to the demand for electricity.

Larger and more efficient power plants were built to supply the increasing electricity demand at a rate which meant that it doubled in eight years. During that time the development of power generation sought to improve the efficiency of the process and raise the transmission voltage.

There was concern over pollution from coal-fired plants which led to the introduction of electrostatic precipitators to collect dust carried over in the flue gases. The height of the stacks, up to 200 metres on the largest stations, would ensure dispersal of the flue gases on the prevailing wind.

Nuclear power had started within this general development of electricity generation that was in progress up to 1970. The first nuclear power station is generally considered to be Calder Hall in northwest England, which was officially opened by HM Queen Elizabeth II in October 1957 and was closed down after 45 years of operation, in 2002. The first Pressurized Water Reactor (PWR) followed at Shippingport, PA, at the end of that year. The 68 MW unit ran for 25 years and was shut down at the end of 1982.

The twentieth century was also a time of social change and not just in greater sexual freedom, but in the questioning of the direction in which society was moving. If there is one event which started the Green movement, it was surely at Christmas 1968, with a dramatically emotional broadcast from the three-man crew of the Apollo 8 spacecraft who were then the first people to be orbiting the Moon. They chose to read the opening verses of Genesis, the first book of the Bible, to underline what they were seeing: the planet Earth, a blue and white ball up in the lunar sky, which was the home of all mankind, and the only one that we had.

This broadcast set people talking about the environment and what effect their activities might have on the world. It was first realized that growth in demand for energy, if dependent on finite deposits of fossil fuels, could not continue indefinitely. It was then not long before new infrastructure developments such as a power station, motorway, or a large industrial site had to present an environmental impact statement as an integral part of the planning process. This would look at the use of the land, what effect it would have on the surrounding communities, and whether it threatened a wildlife habitat; did it impinge on an area of outstanding natural beauty, what sort of wastes did it produce, and how would it dispose of them.

Five years later, in the autumn of 1973 war broke out in the Middle East between Israel and its Arab neighbors, Egypt and Syria, who sought to recover land taken from them in 1967. At the same time the Organization of Petroleum Exporting Countries imposed a four-fold increase on the price of a barrel of oil.

At the time there were a large number of oil-fired power plants in Europe and much of the rest of the world which had suddenly become very expensive to operate. But since most of the oil used in Europe, and increasingly in North America, came from the Middle East, something had to be done. The other big market, transport, was also paying more for its fuel, and this was an easier target for governments to address. Speed limits were introduced and various other short-term measures to limit car use and supposedly to cut oil imports.

The crisis had three long-term effects. First there was a concerted effort to find new fuel resources outside the orbit of OPEC. Immediately this meant development of the oil and gas fields, in the North Sea and Alaska, while elsewhere gas fields were being discovered and developed in the Indian Ocean off Mumbai, in the Gulf of Thailand, in South America, in and around Australia, and in the northern North Sea off Norway. Gas was coming into Europe from North Africa and Siberia, but as natural gas entered the market it first replaced coal gas in the domestic market. In Europe at least, natural gas did not become a mainstream fuel for power generation until about 1990.

Second, a large nuclear power programme had, by the end of the century, installed some 450 units around the world. Some of the earliest nuclear plants have now been closed, but there are still 439 in operation and 35 under construction, including three in Europe and one in the United States.

Work on the development of more efficient energy systems led to the introduction of larger coal-fired power plants with supercritical steam conditions and compulsory environmental additions of Flue Gas Desulphurization (FGD), and more efficient burners to further reduce emissions. Larger gas turbines running at synchronous speed led the development of the gas-fired combined cycle to the point where, since 1990, over much of the world, it has become the preferred choice of power plant for system expansion.

Third was the arrival on the scene of a global protest movement which began to attract public attention. The Green argument was that governments had got us into this mess in 1973, and we couldn't trust them to find the answers.

The first nuclear power plants had come into service and were seen to be reliable in operation and with low fuel costs. Furthermore the uranium fuel came from politically friendly countries, mainly Australia, Canada, and the United States. By 1973 there were forty seven nuclear power reactors in operation in Europe, twenty eight in the United States, five in Canada and five in Japan. But a look at the records of nuclear power in these areas shows that governments had already made plans for more nuclear stations, and a large number went into operation up to 1990. Construction of some of these, having started before 1973, it cannot be considered to be a reaction to the oil crisis as such.

Although the Green movements came to the fore in the early 1970's they were first interested in the influence of energy technology on people and the environment. Los Angeles was known around the world as a city totally dependent on cars, and with an almost permanent smog in the daytime. But it was not the only city so afflicted: flying around the United States at the time, as the aircraft descended there would come a point where it would suddenly shudder as it passed through a temperature inversion.

There began a series of actions which aimed at improving the environment and human health particularly in the cities. Fifty years ago lead tetra-ethyl was used as an anti-knock agent in gasoline until it was claimed that airborne lead from car exhausts in the city environment could retard brain development in children.

After 1970, at about the same time catalysts were developed to remove nitrogen oxides from vehicle exhausts, which were claimed to be responsible for increased incidence of asthma in the population, and because lead would poison the catalyst, unleaded gasoline was introduced first in the United States and later in the rest of the world.

Acid rain also began to be noticed after 1970 and was attributed to the presence of sulphur in coal, which when burned in power station boilers would produce sulphur trioxide, which on contact with moisture in the air would form sulphuric acid. Sweden, in particular, complained of the acidification of lakes in the south of the country, which were downwind of large coal-fired power plants in Denmark and the UK.

Many of the older plants which had originally been designed for coal firing had later been converted to burn oil, which often had relatively high sulphur content and many of these older, less efficient power plants would be replaced by the new nuclear and coal-fired stations. So the first move was to develop FGD systems and fit them to all new coal-fired plants. The first installations appeared at the end of the decade on power plants in Germany and the United States. Unlike coal, those power plants which still burned oil could have the sulphur content reduced in the refining process.

So there was the beginning of an environmental clean up which has continued to the present time, and which has not significantly altered our way of life. But it has not proceeded as fast as it might have done. Improved combustion systems and exhaust cleaning have resulted in the reduction of nitrogen and sulphur oxides from power plants and cars.

Although the efficiency of power generation has improved, there are still in service many coal and oil-fired power plants from earlier times, which have lower steam conditions and few if any of the environmental measures which are now required for all new coal-fired plants. The combined cycle, gas-fired, with high efficiency, and low environmental impact, was by the end of the century accounting for the majority of capacity additions over much of the world, and particularly in the developing countries of southeast Asia

In 1975 only 30 years had passed since the bombing of Hiroshima and Nagasaki had brought to an abrupt end the Second World War in the Far East. There were many people alive then who would remember newsreels, and had read reports of the aftermath of these events, and enough were ready to believe from a position of ignorance that if anything went wrong in a nuclear power station there would be similarly a huge explosion that would kill them all if they lived anywhere near it; after all the fuel was the same material that had beeen used in the atomic bombs

But there was another issue which was even more potent and that was the use of plutonium. This is a transuranic element which does not occur naturally. It is a product of the nuclear reaction and reprocessing separates it from the spent fuel. The only thing that is widely known about it is that it was the material used for the Nagasaki bomb, and that one of its isotopes has a half life of 24 000 years. Yet it is a valuable fuel material in its own right, is responsible for about one third of the output of any nuclear power station, and has been used in the generating units for the long range spacecraft to the distant planets and beyond.

The misrepresentation of plutonium was put at the heart of the anti-nuclear case. We could not have a fast breeder reactor with a plutonium fuel cycle and we could not have mixed oxide fuel in the present reactors, because in both cases more plutonium would be produced. It must not be allowed to develop an industry based on bomb material because it could not be trusted to keep a proper inventory of all the material and account for any losses that might occur. Could not some of the material be leaked out to terrorists who could make their own bomb?

Opposition to nuclear energy grew first in the United States where there were plans to build another eighty reactors, to replace oil-fired power stations which were mainly in the southern and western states. Growth in electricity demand had slowed down and Green opposition at the various licensing stages had managed to extend the proceedings so much that the costs of the hearings were getting out of hand and many applications were abandoned.

Green protest had been seen to work, and the increasing presence of the leaders on the news media who regarded them as expert commentators on nuclear technology and development plans, ensured that there would be public exposure to their views. Their particular view on energy was that conservation would reduce demand which could then be met by renewables. But this is a contradiction in terms. There is nothing wrong with conservation in terms of having adequate insulation and more efficient electric machines and lighting; this has been at the heart of industrial development over the years.

It is the renewable energy which is the problem: large numbers of small units spread out over a large area. The heavy carbon footprint of production of all this equipment, and the much greater demand for steel and copper compared with a traditional thermal power station of the same output, has been conveniently overlooked by the advocates of renewables. But in reality what this view advocates is the construction of generating plant with no emissions, but unable to meet the demand placed on it for 24 hours a day and 365 days a year.

A consequence of Green activity has been a general extension of the planning process for all large infrastructure developments which has now got to the point that governments are starting to react to change the law governing the licensing of important national projects and curb the political infiltration which has seen many, particularly

European countries, adopt the Green energy prejudices. The result is less construction of new plant and great proclamations of faith in renewables.

Awareness of global warming developed at a time in some countries of government by one party that had been in power over several terms for a long time: Conservatives in the UK, Christian Democrats in Germany, and Republicans in the United States. Opposition parties in desperation sought support from groups of single-issue fanatics who opposed a particular government policy and found in the Greens willing supporters.

Thirty years after the 1973 oil shock, the world is again at a time when we are considering the way we generate electricity. Then the main concern was to take oil out of power generation. Today it is the removal of carbon dioxide out of flue gases and the introduction of other carbon-free technologies.

In about 1990 global warming started to come to public attention. At least much was said but very little was done. In the beginning the question was whether it was due to human activity or a natural phenomenon. Global warming has happened in the past and been followed by cold periods which suggests that it may have more to do with perturbations of the earth's orbit around the sun than any activity on earth. But if this is happening again then there is a ready explanation.

In the sixty years since the end of the Second World War, global population has trebled from 2 to 6 billion. It has happened in a period marked by great economic development, not only in industry but in agriculture and medicine. People are living longer and are eating better than in the past. All of this has led to growth in the production of electricity, and in the transport of goods and people around the world.

In 2000 there were 172 nuclear power plants operating in the then European Union. If these had not been built, coal- and gas-fired power stations would be supplying this energy, and emitting 500 million t/year of greenhouse gases. For the 450 operating plants in the world, then the savings of emissions would be about 1.4 billion t/year.

But while these plants were being built between 1970 and 2003 the population doubled. We live on a sphere approximately 12800 km in diameter covered in an atmosphere more than 30 km thick, which is a lot of gas and almost everything that we do happens in the bottom 10 km of it. The claim for global warming is that carbon dioxide and other greenhouse gases collect in the upper atmosphere and trap heat which would otherwise be reflected into space. This might explain a succession of warmer summers and other unusual weather patterns.

The two forms of life on earth, animals and plants, play complimentary roles in sustaining each other. Animals inhale oxygen and exhale carbon dioxide which plants absorb and give out oxygen. Carbon dioxide emitted close to the ground by people and car exhausts is surely absorbed by trees and plants around us.

Global warming can be said to be the result of growth in population, and the increased use of energy that results from it. But we must be careful not to confuse two distinct issues: natural changes in global climate resulting from perturbations of the earth's orbit causing it to move nearer to or further from the sun, volcanic eruptions, sunspots, or other long period cyclic effects; and emissions of greenhouse gases from combustion of fossil fuels, and the increased poulation.

Emissions can be controlled or eliminated, and it is desirable that they should be. Burning coal or oil to produce process steam or to generate electricity has resulted in factories and power plants having tall chimnies to carry the smoke away on the prevailing wind.

Nuclear power came on the scene when there was growing concern over the efficiency of electricity generation at a time of rapidly growing demand. The completion of Calder Hall solved one of the problems in that it was 200 MW of generating capacity with no emissions to affect public health. It had come into operation two years after the British Govenment passed the Clean Air Act and four years before the first combined cycle was completed at Korneuburg, Austria.

But with the oil crisis, public opinion became more outspoken as measures were introduced to reduce consumption of oil. The environmental concerns which had developed at the end of the sixties were more in evidence, and confidence in government was diminishing.

While the first technology protests were environmental, it was the protests against nuclear power plans which came to the fore at the end of the 1970's and by 1990 had all but brought nuclear construction to a halt in North America and Western Europe, but not in the rest of the world, and notably in China, Japan, Korea, and Taiwan.

The Green argument was that we were wasting energy and therefore did not need to use so much. Were these new power plants really necessary? Should we not do more to insulate our homes, and turn down the thermostats and drive smaller cars. But even when you have done all that you still need electricity for lighting and cooking and the industrial controls and processes, so power plants have still to be built.

During this period there were many countries around the world with monolithic, state-owned, electric power utilities who saw it as their duty to provide an affordable and reliable supply of electricity to everybody. In Europe that was the Central Electricity Generating Board in the UK, Electricité de France in France and ENEL in Italy, who although they were perceived as leaders in electricity generation and transmission technologies turned their backs on combined heat and power, the one measure that would have contributed to a significant energy saving.

In northern Europe, particularly Scandinavia, Germany and the Netherlands, district heating had been developed in many cities based initially on coal- and oil-fired steam turbines and later gas turbines. Sweden in the early 1970's even proposed nuclear plants for district heating, but it was not followed through. But they did replace some oil-fired package boilers on their district heating networks with electric heat pumps.

The objection to any form of combined heat and power was that it reduced the production of electricity, which was the prime purpose of the utility. If they bled steam off the turbine for district heating or industrial process supply, less electricity would be produced and they could not earn as much from the sale of heat as they could from the sale of electricity.

There was no such objection to a gas turbine with a heat recovery boiler. So it was no surprise that the United States took the first step with the Public Utilities Regulatory Powers Act (PURPA) of 1979. This created a market for combined heat and power which, since it was largely gas turbine based, played an important role in the development of gas turbines and low emission combustor systems across the industry. But it also promoted energy economy through combined heat and power, at least for industries which were naturally big energy users.

Why not renewable energy? These were all options being considered. Renewables, of which wind power is the most widely developed of the new technologies, are small scale systems and not available all of the time. Tidal energy, is only available for a few hours either side of high tide, and hence the availability of the power plant changes as the moon rotates around the Earth. Solar plants only work during the daylight hours, which in more northerly latitudes varies with the time of year. Wind generators, both on and off shore, can only operate within a range of critical wind speeds, which can occur at any time.

If other countries had followed the PURPA legislation in the United States it might have been different because what PURPA did was define what industrial electricity generators could do and also how they could trade electricity at a fair price. Although is was not obvious at the time it effectively started the separation of transmission and distribution from generation, and for a large number of combined heat and power schemes this reduced the cost and increased the efficiency of energy production in a wide range of industrial processes.

When, ten years later, privatization of electricity supply in the UK created the functional separation of generation and transmission, it kick-started a global expansion in combined heat and power. If anybody could generate electricity and guarantee availability, and sell it at a fair maket price, then how it was generated was irrelevant. Investors wanted a plant which was quick to build and operate, and could offer a rapid rate of return on the investment; which all pointed to the combined cycle, with its speed of construction, low environmental impact, and high thermal efficiency.

All during the 1980's, as Green influence infiltrated public opinion, the electricity generators saw no future in, particularly, wind power which was expensive to build and intermittent in operation. They could improve efficiency and cut emissions of thermal plants, but wind had a bigger environmental impact with dozens of units of low capacity.

The object of the Kyoto conference at the end of 1997 was to discuss how global warming could be reduced through cutting emissions from electricity production and energy use, and developing environmentally friendly systems of electricity generation. This was considered to be so important that targets should be applied to define progress, which were set up in a follow-up conference, in the Netherlands in 2001. This set targets for emission reductions to 1990 levels by 2012, when a second stage would be defined, with more countries, to carry through to 2020.

It was this conference which effectively pushed the Green Movement's agenda in support of renewable energy. By 1997 the deregulation of electricity supply around the world was well advanced, which set the conditions for small generating sets to be installed by industry and the new generating companies. Wind farms started to appear across Europe and North America, initially units of up to 2 MW on land, and later in response to public objections, at 3.7 MW and upwards offshore.

The countries in the initial phase of Kyoto are chiefly in Europe and North America, allegedly the most polluting countries, which should show the rest of the world what to do. But this is a relative term, the Green activists are fond of pointing to the United States as the biggest polluter of the lot. Yet for more than forty years that country has worked hard to improve its own environment and received no credit for it at all. The result is that across the country there has been a marked reduction in emissions, particularly of sulphur and nitrogen oxides both from vehicles and power plants.

The United States has always relied on technology to reduce emissions, and it is technology which will reduce electricity demand. At present the nuclear fleet is having its steam turbines upgraded and an increasing number of stations are getting the license extensions to sixty years. If each one of 104 operating reactors receives an upgrade of 50 MW, that is equivalent to five additional reactors being added to the system. Four new reactor designs have been licensed by the Department of Energy, and several generating companies are now actively planning new nuclear plants for service after 2015. The Electric Power Research Institute (EPRI) is even looking at what would be needed to license the existing nuclear plants for longer than sixty years.

There is a continuing market for wind generators across the 48 contiguous States. But it is not for this alone that emissions are falling. The American view was that targets and taxes, were not the way to produce results and it seems to have had some effect. Net greenhouse gas emissions fell by 1.5% from 2005 to 2006.

The European Large Combustion Plant Directive (LCPD) which was announced in 2001 requires all coal- and oil-fired power plants built before 1987 to be shut down by the end of 2015, unless they have opted in by fitting FGD systems and other environmental measures before the end of 2007.

LCPD is among the more sensible post-Kyoto decisions which has been accepted by the now 27 European Union countries. which are required to achieve a 20% cut in greenhouse gas emissions by 2020. It amounts to a timetable for cutting pollution, because more efficient and environmentally friendly power plants that replace the old stations will have lower and cleaner emissions, and must be of an equivalent capacity. If not, then severe power cuts may follow, which will require serious load shedding day after day to control the frequency.

So before we consider the energy system of the future, if we are to use less then it is down to us as individuals to play our part. Low-energy fluorescent lights are gradually replacing incandescent bulbs, but more for the sake of economy. An 18W low energy unit costs more but will give out as much light as a 100W incandescent bulb and last up to ten years. Around the world utilities and some supermarkets have been offering attractive prices for bulk purchases of the fluorescent units.

If a housholder with 100 W light bulbs, in each of the five major rooms of a typical three-bedroom house, were to change only these bulbs for low energy units, the load would drop by 410 W. Repeat this in a million homes and the total reduction of 410 MW is the equivalent capacity of a single-shaft combined cycle plant in the 50 Hz market.

Similarly if there is the chance to generate some electricity from a photovoltaic array on a sun facing roof, for as many as can afford the installation there are those who don't understand the point of doing it. If a government offers grants to encourage this as an issue of energy policy, there are those who will take it up as of right, and those who do not want the hassle of applying for the grant and imagine that all sorts of objections will be raised which will delay or prevent it happening.

But how do we define a green energy system? First, it must have minimum impact on the environment, in terms of its emissions and would not have a hidden component of environmental damage through excessive energy loads in the production of materials, manufacture of the components, and the erection of the plant.

This in fact is the reason why, since the deregulation of electricity supply, in many countries the combined cycle has become the preferred option for new generating capacity. Environmental impact includes such things as physical size, traffic levels going to and from the operating plant, noise levels at the site boundary, and impact on cooling water. Screening of the station by tree planting can also contribute to reducing the environmental impact.

But the technology must also be sustainable in as far as it can guarantee to the generations that follow us a secure energy supply at an affordable price. People 100 years from now may have a somewhat different way of life compared to ours, but they will still need energy to support it, and electricity demand could be much higher than it is now.

There cannot be totally emission-free electricity supply because it is generated to meet demand as it happens. Even if all the base load is carried on nuclear power plants there will still be a need for thermal plants of great flexibility which will be always available to cover nuclear refueling and maintenance outages, lack of rainfall causing low hydro output, and of course the low availability of some of the renewables, as well as daily peaks in demand.

But at the end of 2008 the global finacial crisis has pushed global warming into the background. So the question is how far can the replacement of old power power plants with modern efficient and environmentally friendly systems be the mechanism for reigniting industrial growth to pull us out of recession?

2
Global warming

lobal warming, or climate change, as a political and environmental issue has, in the last twenty years, been pushed to the forefront of public opinion. While there are many examples of environmental change which could be attributed to a warming trend, there is no accurate account of similar happenings in the past when the world population was much smaller and water, wind, sun and biomass were the energies used

Green alarmists would have us believe that if we don't do something quickly, then the small low-laying Pacific islands and large coastal cities such as London, Amsterdam, Boston and New York would disappear under tens of metres of seawater before the end of the century.

For the time being the global economic crisis has, since mid 2008, pushed the issue into the background. It is now the proceedings of learned societies that hold the current views on the subject which is that it is happening faster than we previously thought.

Climate and change are now the buzz words of the Green movement. They want to strike fear into the public just as they did over nuclear energy, in the forlorn hope that this will lead to a future world order of simple living with lower energy demand. The underlying argument is that the more carbon dioxide that we pump into the atmosphere as the exhaust of the combustion process in power plants, boilers, motor vehicles, ships and aircraft the concentration in the atmosphere increases. It then rises to the top of the atmosphere where it will block radiation of heat back into space and raise global temperature in a greenhouse effect.

Since the basic mechanism of global warming is excessive discharge of carbon dioxide into the atmosphere, there is one source which cannot be controlled which has continuously increased. The population of the world has more than trebled since the end of the Second World War in

1945 and is now about 6.5 billion with a parallel increase in the number of animals for food, for working, and as pets. All of them breathe in oxygen and exhale carbon dioxide. Three times as many people require three times as much water, and three times as much food. They require somewhere to live and their need for energy in their homes and at places of work has over the years produced growing demand for energy to support greater economic activity.

Climate change would be a consequence of serious global warming. A sudden increase in concentration of carbon dioxide in the atmosphere has been observed in the last twenty years which has given some credence to the theory, particularly as unusual climatic phenomena are observed. These could be a period of years with seemingly very warm summers, or heavy rainfall at unusual times of year. The arctic ice sheet is decreasing with the result that a sea passage has been opened across the top of Canada: the famed Northwest Passage.

In reality, the only things of which we can be absolutely sure are that the global population is continuing to increase, and that we are in addition continuing to pour carbon dioxide into the atmosphere from our cars, aircraft, ships, factories and power stations. The present economic situation has temporarily reduced some of these activities and may continue for some time.

What we do not know is what happened at other times of climate change in the past. We know that there was a Viking settlement on the south coast of Greenland in mediaeval times, which died out in the fifteenth century when a falling temperature made it impossible to continue farming there. The continuation of this cold spell for the next four hundred years meant that in England it was possible to skate on the Thames in winter during the mid-seventeenth century.

Looking back further, in Roman times it was possible to grow grapes in England and make wine. Commercial wine production has only started again in England since about 1970. So it would seem that the world has experienced a cold period from at least the end of the fourteenth century up to the mid-nineteenth. At that time the population of the Earth was little more than 500 million and energy was gained from wind, biomass, and animal power until the industrial revolution.

This would suggest that these were large climatic changes due to perturbation of the earth's orbit around the sun, since these were on a global scale.

With the industrial revolution there began mechanical transport by steam, both in ships and on the railways. Later the internal combustion engines, and finally the gas turbine, extended this to road transport and aviation. All these engines started as relatively inefficient devices but over time improvements were made, resulting in higher performance with greater fuel efficiency for increasing numbers in use.

The big change at the end of the nineteenth century was the start of the commercial generation of electricity, which again has made great strides in improving efficiency. But all the while this was going on the global population has increased. Thus as demand for energy has increased so the amount of carbon dioxide discharge has increased.

This is undeniable; the global population is now more than 6.5 billion, of which more than one third is in just two rapidly developing countries: China and India. The mega-cities of Asia and South America, places such as Shanghai, China, Manila, Philippines and Sao Paolo, Brazil, are further concentrations of urban population of high energy density; the result of rural populations moving to where there might be work now that farm mechanization has denied them opportunities in the countryside.

The result must be that as the definitive energy-intensive life-style of Europe and North America spreads around the world, demand for energy will continue to increase in total while demand for energy in particular activities will decline through application of new technology of greater efficiency. But as in Europe, higher living standards and longer life expectancy may eventually see a lowering of the rate of growth of the total world population.

The challenge, therefore, must be to create a reliable and efficient electricity supply system as a first step because of all energy systems it is concentrated on a large scale. The relatively small number of power stations in each country makes it easy to legislate for particular environmental measures as has already happened. The future electricity supply system which evolves should have minimum environmental impact both in its construction and operation and, above all else, the electricity produced must be affordable to all users.

The irony of the present situation is that some fifty years ago, the low efficiency of thermal generating plant was a matter of public concern at a time of rising electricity demand. The solutions found at the time are still with us and there has been a continuous improvement in performance and environmental impact of power plants in the years since.

The other energy use to be changed is transport which is more difficult. Because of the greater number of units the manufacturing base must be changed, the fuel must be changed and introduced to market while the current system is run down. The likelihood is that for road vehicles, at least, a possible solution is the use of hydrogen powering a fuel cell driving electric motors. This will require power stations specifically to generate the hydrogen and liquefy it. The alternative is battery power which would require a heavy charging load depending on use of the vehicles.

The building blocks of the new electricity supply system could be: ultra-supercritical steam plant with efficiency of 45% compared with 36% in the best subcritical units; combined cycle with efficiency of 60% with the new gas turbines now becoming available; nuclear systems both water-cooled and with the new, smaller gas-cooled reactors extending the range of applications which would be completely free of emissions. Then there are the renewables: classic hydro and pumped storage systems which are fully integrated with the present electricity supply system.

The new renewables, wind, tidal current, and solar, are not significant in their contribution to present energy supply and have a number of drawbacks. They are small generating units; the largest wind generator is only 6 MW and the largest tidal current unit is 1.2 MW. Wind farms at least, have a very large environmental impact because they are so highly visible.

Solar photovoltaic power is potentially the most important of the renewable technologies. The active cells are solid state electronics which can be expected to reduce in price as production increases. It has the potential over a large part of the world to bring electricity generation down to the consumer level.

A photovoltaic array applied to the south-facing roof of every house would supply up to 4 kW to each household during daylight hours and any energy not used would be sold to the grid. For this to happen the house must have an intelligent meter which would indicate the power imports from and exports to the grid, but also there must be enabling legislation to stimulate the market.

These are the tools available for the future electricity supply system but are they simply replacing old and less efficient plant with new and more environmentally friendly systems, as electricity demand continues to grow; or is it to curb global warming. But it also requires enabling legislation to achieve full use of the various technologies.

The practical reasons for building a new power station are to add capacity to meet increased demand and generally the new plant will be the more technically advanced. A supercritical coal-fired plant at an efficiency of 45% will burn more than 20% less coal than the old plant it would replace. Its generating costs will be less because of its lower capital and fuel costs.

Two 800 MW units will replace three 500 MW sets. The generators will be more efficient and the steam turbines will also be of a simpler 4-cylinder design rather than a larger version of the old 5-cylinder sets, and the blading will be of computer-aided design and consequently more efficient. So the resulting plant will be more efficient and more economical to build. The same principles will determine the design of the turbines of combined cycles and nuclear plants.

But the type of plant to be built may not be entirely the choice of the owner. Governments still decide what will be built and when, and it is here where the global warming issue has taken hold. Even though investor groups may build and own the power plants and control the transmission and distribution system, it is governments that give consent for power plants to be built.

This is where the Green movement has entered the fray with its no nuclear, all renewables and conservation message, which over the years has found willing listeners particularly among opposition parties who have been out of power for a long time. Where they have come back into power they have been willing advocates of renewable energy, and many holding the view that nuclear energy if it was to be used at all would be a system of last resort.

The big climate change event was the Kyoto Conference at the end of 1997. The more time that passes it would seem that this was a Green-inspired event by the way that countries have responded to the conclusions. The developed countries of the world would apply measures to meet a target of emission levels reduced to those of 1990. The follow-up conference in Den Haag in 2001 was the defining act. All countries which ratified it would be bound by the targets set for emission reduction.

But the problem is political. The last twelve years has seen changes in the political make-up of some European governments, May 1997 saw the election of a Labour Government in the UK after 18 years in opposition. Gas was already taking over electricity supply, but six months after their return to power, during the Kyoto conference, the British government announced a ban on new gas-fired generation, no doubt to protect a declining coal industry.

Yet since then the only power plants to have been built in the UK are gas-fired combined cycles, along with a few small hydro schemes and a number of offshore wind farms. Much has been said about future energy policy but very little has been done. A number of old coal-fired power stations have been pulled down and are being replaced with gas-fired combined cycles.

In Germany 1997 also saw the election of a new Federal Government after a long spell in opposition. The Social Democratic/Green Party coalition, which had in earlier State governments refused licenses for new nuclear power plants, were determined to shut down all 21 operating units which supplied 33% of the country's electricity. The nuclear industry managed to calm the situation down and the plants will continue running to the end of their design life.

Meanwhile the country has been plastered with more than 20 000 MW of wind generators, each of less than 2 MW capacity, which are now attracting public protest along with plans for new coal-fired plants and, as in the rest of Europe, gas-fired combined cycles are being built as the only acceptable option.

In the sixties the generation of children that were born in the years immediately after the Second World War were entering adult life, with some going to universities and others entering the job market. The shock of the Kennedy assassination in November 1963 had been followed by the triumphant spectacle of the first landing on the Moon in July 1969.

In hindsight, it was the moon landings which were undoubtedly the event which started people thinking about the environment. By the time the Apollo program finished, some thirty people had either orbited or landed on the Moon and had the same view of a blue and white ball up in the lunar sky; the planet Earth, the home of all humanity. But on it there were two ideologically opposed power blocks with enough weapons to destroy each other and everybody else on the planet as well!

The sixties had been a period of economic growth, with full employment, and rapid improvement of living standards, which were reflected in a steady growth in demand for electricity. But in the late 1960's the combination of student unrest and a flood of liberalizing legislation on sexual and racial issues created a situation in which, with the new found freedom of expression, signs of protest at the pace of change began to appear.

That it happened in first in North America was inevitable because protest can only be effective in a free society where any member of the public so motivated can write to a newspaper or contribute to a radio or television programme. This was never more so than in the United States and Canada where centers of population are so widely spread, and across several time zones, that much of the news media are locally based. In such an environment it is easy to get one's views heard and particularly if it touches on contentious local issues such as the route of a new highway or transmission line or the site of a new airport.

When, in the autumn of 1973, war broke out in the Middle East and the Organization of Petroleum Exporting Countries imposed a 4-fold increase in the price of oil, the response of the developed world was to reduce dependence on oil, at least for power generation.

Developments over the previous twenty years had shown how electricity could be generated by a clean technology with no emissions to acidify lakes in another country downwind and hundreds of miles away. A method had been found to transmit power economically over long distances and without propagating a fault from the sending network to the receiving network. The results would be fewer emissions and greater security of supply.

There were no other fuel options than nuclear, coal, and whatever water power could be developed. Some of the big gas fields around the world had been discovered but development was in its early stages and the extent of their reserves was not fully known. The first high power, industrial-frame gas turbines capable of running at synchronous speeds were just starting to appear and the first combined cycles to incorporate them had appeared in the United States, but in Europe the initial priority for natural gas was to replace coal gas in the domestic market.

For the developed countries of Europe and North America with their high levels of demand for electricity the problem was to remove the base-loaded oil-fired plants from power generation, which meant replacing them with coal or nuclear. But not all was well with public opinion in much of North America and Europe. There was a feeling that government had become remote.

These were the middle years of the Cold War. Up to then, anti-nuclear protest had been political in origin, but all that was about to change. Protest had already appeared in the context of high voltage overhead lines and pipeline routes. The towers in their various designs, which got bigger with the voltage, and the bundles of cables suspended from large insulator strings, were considered by many to be an eyesore, spoiling the landscape and potentially harmful to people and animals living underneath them.

But what if the animal or even a plant living under the power line or in or beside a river, was a unique species: real or imagined? This was the next challenge. Build a dam on that river and you will destroy the habitat of a unique fish species, that nobody has ever heard of before, or the reservoir will flood the only known place where a particular plant is found or destroy an ancient historical site.

With these first protests emerged the now well known environmental movements, such as Sierra Club, Friends of the Earth, and Greenpeace which were starting to preach a different philosophy. We did not need all these new power stations, we could save energy by better insulation and improving efficiency and wouldn't the world be a much better place if we could produce the energy we needed from renewable resources?

This was a protest against technology, but it was also a call for the protection and preservation of the natural environment. The first change in response to this was that any infrastructure development such as a new motorway or power plant would require the production of an environmental impact statement. If it was accepted, then the project could go ahead, but this did not mean that it would be free from protest during its construction.

The period after 1973 has brought a wide change in society in their relationships with government and the scientific community. Science has become more specialized and less easily understandable by the public at large. The big nationwide protests had to touch on issues of public concern for them to get any following or any widespread acceptance of the results.

If we look at the actions which resulted from organized protest there is a common thread running through them and that is fear of the unknown. For unleaded gasoline there was the fear that if we lived in towns with dense traffic our children would go mad if they breathed in lead from the car exhausts. For the ozone holes above the Arctic regions, would we get skin cancer sunbathing on our holiday beaches because of the ultra violet radiation that would flood in.

We didn't want nuclear energy, lest something might go wrong in the power station which would cause it to explode like a nuclear bomb. This has never happened, not even at Chernobyl, in 1987, where fewer than 100 were killed and the three undamaged reactors were later returned to service. If the effect had been that of a nuclear bomb the city of Kiev might not now exist.

The most long lasting and widespread result of protest has not been on nuclear energy but on the more widely used technologies of transport and agriculture: the phasing out of leaded gasoline and the fitting of exhaust catalysts on all vehicles.

That was followed by a general protest against big infrastructure developments. Electric utilities have for many years complained of the difficulty in finding sites for new power stations, regardless of type. For nuclear power plants and related facilities there was ample opportunity to challenge at every step of the licensing procedure.

Green activists, particularly in the United States had a field day. The protracted hearings forced design changes to reduce the risk of impossible accidents, which added to the costs, to the point that many utilities threw in the towel and cancelled many of the planned power stations. This however did not have any serious consequences because of the lowering of the rate of growth in demand for electricity at the time.

Despite this, by 1990 the nuclear power plants in Belgium, Finland, France, Germany, Netherlands, Spain, Sweden, Switzerland, and the United Kingdom were nearly all complete and in operation, with only the 1100 MW Sizewell B in the United Kingdom still under construction. The collective result was that in the then European Union more than 30% of the electricity supply was provided by nuclear power plants with no emissions of sulphur or nitrogen oxides, and in total more than 500 million tons/year of carbon dioxide was not discharged into the atmosphere, but would have been if instead fossil fuels had been burnt to supply the equivalent output of electricity.

Logically any government which was concerned about global warming would create an electricity supply system with minimum emissions and encourage greater use of electricity where possible in industry. But that was not to be. The Green Movement had effectively put a stop to nuclear plant construction in the United States and much of Europe where it had successfully infiltrated the political parties and influenced public opinion, and it was widely felt that no more would ever be built there.

What we are seeing now is a world run by technology and governed by people who don't know the first thing about it. It is a situation which has gradually developed over the years as Green opinions have infiltrated the teaching profession with the result that fewer students are entering university to study science, mathematics or engineering.

So Governments have to hold a balance between, on the one hand their electorate, who have been exposed to all the Green opinions and, who firmly believe that global warming is happening and that dire problems will result if we don't do something quickly; and the Scientific community, on the other, who seek to find out what is actually happening and to determine what could be done about it, if anything.

We hear for example that glaciers in Greenland and the Antarctic are melting faster than in the past; but since how long ago? We only know what happened during past climate change through the writings of the time, which were essentially the observations in the immediate vicinity of the author, without any firm scientific evidence to back it up. Five hundred years ago the world was still opening up to exploration, and when it took the better part of a month to cross the Atlantic it took a very long time for communication between the continents.

Today Greenland is as warm as in the time of the first Viking settlements. The sea channel between the west coast and Baffin Island is free of ice all year round and the majority of communities are on that side of the island. The only ice normally seen is icebergs drifting south, which have broken off glaciers in the far north.

Kyoto came at a time of increasing influence by the Greens on government and the increasing number of people in government who were almost professional politicians with minimal experience of industry or the law. Several countries initially refused to ratify, but these were gradually whittled down to just the United States, whose objections were always that technology, not targets supported by taxation, would cut emissions, and that the original Kyoto Protocol did not include the fast developing Asian economies, particularly China and India, with their rapidly increasing demand for energy, and electricity in particular.

At the end of 2007 a United Nations conference in Bali brought 190 countries together to thrash out a global plan for combating climate change. Again it was the United States against the rest with the European countries wanting a global reduction of 40% by 2020. But the outcome of this is another conference in Copenhagen in 2009 to finalize a plan as to how it should be done and which every participating country will be expected to ratify.

It is not difficult to reduce emissions from the easily controllable industry of electricity generation. In fact the answers are with our present generation and transmission technologies. Transport is a bigger and growing source of emissions and while carbon dioxide emissions at ground level can be absorbed by grasses, trees and plants at the roadside, taxation based on engine capacity or congestion charges in cities, are simply putting extra costs on business which would mean higher costs of transport which would inevitably be passed on to the consumers.

Any binding target for emissions reduction is at the mercy of politicians to implement, and since it might require unpopular decisions, such as the imposition of taxes to deter use of vehicles and aircraft, who then is going to take the plunge with an election approaching which might cause them to lose power.

However the reality of the situation is that although we can calculate how much greenhouse gas is emitted by any fossil-fired boiler, power station, car, ship or aircraft, we have a far from clear understanding of the cumulative effect on mean global temperature, given that it has increased by less than $1°$ C during the 20th century. Furthermore, during that time the efficiency of energy production and end uses has steadily increased.

Electricity generation accounts for 22% of global greenhouse gas emissions; this does not of course include nuclear, hydro and wind power, nor does it include the energy cost of its fuel supply, without which the power stations cannot function. However electricity at the point of use is totally emission-free.

To achieve a large reduction of greenhouse gas emissions, more energy use must be based on electricity which means that the emissions are applied to the power plants. So the new power plants which must supply this increased demand should have either none, or very low emissions.

How can this be done? First we can reduce fossil-energy demand by going to nuclear and renewable technologies and taking coal out of power generation.

Gas compressors on some of the older pipelines are being replaced with electric units, but this is only being done in developed countries with ready access to an electricity supply. The typical gas turbine driver for these early compressors would have an efficiency of between 25 and 30%. The electric motor drive could be powered by a combined cycle at 58% efficiency or from a hydro or nuclear power plant which would have no emissions. Either way, no gas is taken out of the line, to boost the pressure: it all goes to the end users.

So there are a number of applications of existing technology which if applied, not everywhere, but in a sufficient number of places, could achieve a significant reduction in electricity demand or of fuel consumption. The generation of electricity more efficiently is only one part of the equation. The other is the efficiency of transmission.

Losses in transmission have increased as voltages have increased and applications of electricity have diversified. Synchronous or reactive compensation systems can improve the situation on long lightly loaded lines, outside of peak times. But the basic losses stem from the distances at which power stations have been built from load centres.

Large hydro stations can only be built where there is sufficient water resource available. Coal-fired stations are built near the coal mines. High-voltage direct current (HVDC) has connected distant power plants to load centres and connected islands to the mainland. The first HVDC connection was between Sweden and the Baltic Island of Gotland, and was followed by the first of two connections between France and the United Kingdom.

The Nordel Power Exchange was set up between the Scandinavian countries more than 50 years ago, joining countries with large hydro potential to others with none but large coal-fired and nuclear power plants which could work as one system. High Voltage Direct Current links were set up to ensure that faults could not be propagated between the national networks.

Peak times in Finland were an hour earlier, and since it was much further north than Denmark where there was not such a wide variation in daylight hours and temperatures between summer and winter, there were big differences in the pattern of electricity demand between countries and between regions within countries. This determined how electricity could be transferred to satisfy peak demand. Coal-fired plants in Denmark and Finland, and nuclear plants in Sweden and Finland provided the base load and hydro power in Norway, Finland and Sweden provides peak power over the whole region.

To try to curb global warming through electricity generation is acceptable because the alternatives are not. Yet the Green agenda still seems to hold priority although the 2010 targets for renewable energy are unlikely to be met and there are signs of panic in decision making. Yet we know how to generate electricity without emitting greenhouse gases, with nuclear, hydro, geothermal and wind power systems, but the difficulty is bringing these together in a system with fewer transmission losses and which in future may see much greater demand for electricity.

To build for this future it is desirable to cut emissions by using only the most efficient fossil fuel systems. This means phasing coal out of power generation and building gas-fired combined cycle to cover gaps in generation while the new power system is being built. Combined cycle will still be required as a fast-acting system to cover peak demand and the non-availability of renewable systems and nuclear refueling outages.

Yet every signatory to the Kyoto Protocol has set targets for up to 10% of electricity to be supplied by renewables by 2010 and have little or no chance of meeting it. For a wind farm with 30 generators to match the 90% availability of a thermal power plant would require 27 of them to be running 365 days a year at their maximum continuous rating, which is something of which no wind turbine has yet been shown to be capable.

This is the fallacy of the Green argument for renewables and better insulation to supply all our energy needs. It is primarily addressed to the domestic market where it is well within the scope of many householders to insulate their homes and procure double glazing and low-energy lighting. Legislation for building codes can define insulation, standards so that new houses at least, can be built with wall cavity and roof insulation, double glazing and condensing gas heaters to reduce overall energy demand.

There are many other measures that individuals can take to control and reduce their electricity consumption. For instance, if enough people change incandescent bulbs for the equivalent fluorescent low-energy units, it could make a big difference to the lighting load for a whole country. But we cannot reduce global temperatures by this alone.

So far, only two countries, Australia and the United States have legislated to take incandescent bulbs off the market. Look at any American utility website and you will find promotions of low energy lighting and even offers of free low-energy fluorescent units to replace incandescent bulbs. This makes sense because if 5 million units were replaced in every state that represents a 20,500 MW reduction in the electricity load for the United States alone.

The Bali conference took place against a growing body of scientific evidence of a warming trend. Notably in the Arctic regions, glaciers are melting in Greenland and the Antarctic ice cap is eroding at the edges but is this due to the ozone holes which we don't hear about nowadays or a more general warming? In 2007, for the first time in living memory, a clear water channel opened up across the top of Canada between the Atlantic and Pacific oceans: the fabled North West Passage.

Because of all this, it is argued that there might be no snow on the Alps and that some iconic animal species such as the polar bear and the king penguin might be extinct by the end of the century.

This is the problem, we can postulate a global temperature increase of anything up to 5°C but we cannot translate that into a volume of carbon dioxide to produce it and interpret that as so many coal-fired power stations or cars on the streets. Above all, we do not know the total of carbon dioxide exhaled by three billion extra people, because that might explain why emission-free sources of energy have neither halted nor reversed any increase in global temperature.

In fact, if the whole global warming trend is the result of human activity, then it is the consequence of one particular activity which is generally performed in bed. While more than 450 reactors were being built, the population of the world doubled, and may well have increased from the 1960 level by 2050 to more than nine billion.

The one thing that is always left out of the equation is the inexorable growth in the global population. We all breathe in oxygen and exhale carbon dioxide during our entire lifetime, as does every other living creature on the planet. Collectively we are the biggest greenhouse gas producers of the lot. Yet to produce enough food for everybody, tropical rainforests, which provide some of the oxygen in the atmosphere, are cut down in the cause of agriculture which now includes the production of bio-fuels using palm oil, oil-seed rape and similar crops.

A rapid growth of population requires more than energy. Twice as many people require twice as much food, and twice as much water to drink, and to put to all the other uses. Even now we have not been able to cope with the increase There are some developing countries still with large populations of which more than 50% are under the age of 20 and living in extreme poverty and with high death rates from malnutrition. But if a warming trend is the direct result of population growth then either more land must be put under cultivation or else more productive plants must be raised.

We are all breathing out carbon dioxide within two metres of ground level. To suggest that it is all going to drift into the upper atmosphere is not sensible when the grasses, plants and trees around us are surely absorbing it.

A number of factors are coming together to suggest that population growth is as big a problem as climate change. It may be that a slowing of population growth will be the first priority and yet we have only just scratched the surface. The industrial revolution was accompanied by a revolution in agriculture with the mechanization of the threshing process and later the production of tractors to replace horse-drawn ploughs, and the development of the combine harvester.

In Europe and North America the number of people working the land has fallen dramatically in the last century. Families which would typically have up to six or seven children in the latter half of the nineteenth century would see most, if not all of them, for the first time survive into adulthood. Family size has dropped to generally two children and at a European replacement factor of 2.6 many developed countries have seen their birth rate drop below replacement level. Japan, it is said, could see its population fall below 100 million by the year 2050.

Attempts to develop population policies have met with limited success. Back in the 1970's the World Bank funded a number of birth control programmes in Asia. China's one child policy has been only partially successful in the urban environment. The Roman Catholic Church has maintained its resistance to contraception, though its followers in Europe have largely taken the matter into their own hands and had smaller families as living standards have risen.

So the new form of protest is to calculate carbon footprints to try and shame us into action. Go into any supermarket and see how the same foodstuffs are displayed all year round. Depending on the time of year, oranges will be offered in British supermarkets from Israel, Spain, South Africa or the United States. A wine merchant, who fifty years ago would have offered only French wine plus a few bottles from neighbouring European countries, now also offers wines from Australia, New Zealand, South America and the United States. Food and drink are some of the largest commodities transported around the world and within individual countries.

The Green arguments on energy have moved on to transport, and bio fuels are considered to be carbon neutral because they represent the harvesting of a crop, which will grow again the following year and absorb the carbon dioxide from the exhausts of the bio-fuelled cars and aircraft.

Already bio fuel cultivation has had an effect on the agricultural market. The price of wheat doubled between the harvests of 2006 and 2007 and this could be seen in higher prices for bread and pasta. The price of animal feed has gone up because rising living standards increase the demand for meat, particularly in China and India. Every hectare of land which is turned over to produce bio-fuel crops is one less hectare for the production of food crops and the grazing of animals.

Environmental arguments have been going for more than thirty years. One of the first suggestions was by the British scientist, James Lovelock, who advanced the Gaia theory, that the planet Earth was like a living organism which could adjust to the pressures placed upon it. One can argue that it already has.

The fossil fuels that we burn today are the remains of plant and animal species from millions of years ago, almost all of which are now extinct as the result of natural climatic changes over several hundred million years. Human life may be putting such pressure on the earth that it will not be able to support it. If this becomes so then millions may die of starvation, thirst and disease until the situation is corrected, and if global warming encourages certain bacteria and viruses to adapt and multiply it may accelerate the process.

The change of Green tactics to focus on global warming is doubtless the result of another development in space which has greatly added to our knowledge of the Earth. It is only in the last thirty years, since the first weather satellites were launched, that we have been able to observe the world's weather as a whole, and learned much more about the weather patterns and the factors that influence them. Satellite data has provided a much more accurate picture of the weather over large areas, and as a result weather forecasting has become more reliable.

It was satellite data which revealed the hole in the ozone layer over the Antarctic and which led to the 1987 conference in Montreal to look for a solution. It was satellite data which has shown in the autumn of 2007 that enough Arctic ice had melted to open a sea channel across the top of Canada between the Atlantic and Pacific oceans. But has this happened in the past? The Northern Territories of Canada are sparsely populated along the Arctic coast and there is therefore no folkloric account of an open sea passage in earlier times.

Columbus and the other early explorers headed west to find a short cut to the Far East. Instead they found the American continents. But was there a route around either end of these continents? Magellan was the first to find the southern route around Cape Horn into the Pacific Ocean in 1521.

Yet many explorers have set out to find the fabled Northwest Passage over the past 500 years and found their way blocked by ice, and more than one ship has been crushed by ice at its winter moorings. Now that the Northwest Passage has opened, what has caused it? By how much is it due to natural causes; and by how much due to the rising per capita energy consumption of an expanding global population?

There has been a small change in average global temperature yet if it is indicative of a longer term trend and it can be definitely attributed to human activity then there is a case to answer. What we can say is that it was certainly not human happenings which brought a cold period in the Middle Ages that made it possible to skate on the Thames in winter 300 years ago. so can the trend to a warmer climate be similar?

We can only base our assumptions on the data that the satellites provide, but this only tells us what has happened since they have been launched. There are historical records of weather in Europe and North America which also point to warm and cold periods in past times. The Scandinavian names of many places in Greenland point to a mediaeval time of Viking settlement there when the climate was much warmer and farming was possible along the coastal strip of southern Greenland. Other records take account of natural phenomena such as earthquakes and volcanic eruptions, which would appear to have had a far greater influence on the weather than human activity alone.

In 1815, and again in 1883, there were two massive volcanic eruptions in Indonesia: Tambora and Krakatau. Tambora on Sumbawa Island, 300 km east of Bali, was the larger of the two. It erupted 150 km^3 of ash, dust and sulphur into the upper atmosphere causing in 1816 a 3°C drop in temperature in Europe and the New England states in what became known as the year without a summer, marked by widespread crop failure and starvation, and among other weather anomalies, snow in Quebec City in June..

Sixty-eight years later, Krakatau erupted in August 1883. It was a smaller eruption than Tambora, but 25 km³ of dust and ash were released into the upper atmosphere again causing global temperatures to drop. Where it was different was in it being one of a group of small islands in the Sunda Strait between Java and Sumatra formed around the caldera of an earlier volcano thought to have last erupted in 415. The eruption was accompanied by strong earthquakes which generated tsunami that destroyed 165 coastal villages on Java and Sumatra, and killed 36000. The Krakatau eruption was extensively reported and the accounts of these observations laid the foundations for the modern science of vulcanology.

Novarupta in Southern Alaska followed in 1912 with 30 km³ of ejecta and effects felt as far away as North Africa the following week. Then in June 1991, Mount Pinatubo in the Philippines erupted some 100 km northwest of Manila. It ejected 10 km³ of ash and sulphur and brought large quantities of metal ores to the surface.

Given the size of these eruptions and the climatic changes which followed with lower temperatures, it is hard to see how the Green arguments for carbon capture (removal of carbon dioxide from flue gases) can hold up. Volcanic eruptions on the scale of Krakatau would reduce global temperature by more than 1°C in a few weeks.

It is hard to see how installing carbon capture for a 1000 MW power station burning 3 million t/year of coal could have a significant effect, and to how many plants in total would it have to be done, over what period of time, to reduce greenhouse gas emissions enough to achieve a similar reduction of global temperature? All the while this was being done, energy would be consumed in the production of metal, production and assembly of components, transport to site and installation,

So given that if mean global temperature is increasing, is it cosmic in origin or human? If global warming is an event going back to the dawn of the industrial revolution, then the steady improvement in efficiency of new systems replacing old has had little effect, either from reduced emissions or greater use of electricity.

Until 1990 the Greens were still a predominantly anti-nuclear fringe movement starting in the United States and gradually filtering across to Europe which they saw as fertile ground. The Green party as a political entity appeared in many countries but none more successfully than in Germany where they initially got into state politics and could start to impose their own ideas.

In November 1989 the infamous Berlin Wall which had encircled the western districts of the US, British and French occupation zones to separate them from the Soviet controlled German Democratic Republic, was knocked down by public protest. For a few years before mounting public protest had resulted in the removal of Communist Government from Hungary, Poland, and Czechoslovakia, which would later separate into the Czech Republic and Slovakia, all of which are now member States of the European Union.

In 1987 a major nuclear accident at the Chernobyl power station in Ukraine had brought the Western and Soviet nuclear industries together, not least because the Soviet designed reactors were widely felt to have weaker safety systems, and as a result a major upgrading program was set in train to adapt the Russian designed safety systems to Western standards.

After the fall of the Berlin Wall, in the following year, in the Soviet Union, an attempted coup against the President and General Secretary of the Communist party, Mikhail Gorbachev, backfired and the Soviet Union was subsequently transformed into the Commonwealth of Independent States. The Baltic States and the Caucasian republics, and some of the Islamic republics of Central Asia left and became fully independent states, with the three Baltic Republics later joining the European Union.

So after 1990 the Cold War was effectively at an end. But with very little nuclear power construction outside of the Far East the Green lobby seemed to have got its way. But we were beginning to see subtle changes in weather, with higher summer temperatures and warmer winters with corresponding changes observed in animal and plant behavior. If this was a sign of a permanent climate change, how could we cope with it?

The Campaign for Nuclear Disarmament had faded into the background, but the Green movement felt that they had got their way with nuclear power, at least in those countries where they were free to protest. So global warming around which the environmental debate has coalesced became their cause celebre.

It was a simpler project with which to project Green opinions, because everybody is conscious of the weather. On any given day it determines what we wear, or whether we take an umbrella with us when we go out. It will even determine our state of health. People can die from hypothermia if they are caught out in extremely cold weather or without any heating. Old people can also die from over heating in a prolonged hot spell, as happened in the hot summer of 2005 in France.

But what the Green movement seems to have overlooked is that if global warming is attributable to human activity then by our own actions we are in part responsible. As the result of Green antics in the American courts at license hearings up to about 1980 the construction of a further 74 nuclear reactors was abandoned. All of these plants which had a total capacity of 84,500 MW could have been up and running by the year 2000, and avoiding some 845 million t/year of greenhouse gas emissions.

We do not know how far the Green view has influenced governments to not plan for nuclear capacity but what we do know is that many countries now have targets for reducing green house gas emissions and generating 15% of their electricity from renewable sources by 2020. It is unlikely that the targets will be met; but although global warming is not just about electricity generation, the methods employed can make a significant reduction to the emissions causing it.

The global warming campaign gathered momentum after 1990 and the collapse of communism in Eastern Europe and with it the removal of the threat of nuclear war, and without the distraction of more nuclear power plants being built. global warming came into prominence. At the end of 1997 governments gathered in Kyoto, Japan, to discuss the issue, encouraged by the reaction to the Montreal conference of 1987 which banned CFC type refrigerants which were accumulating in the upper atmosphere and damaging the ozone layer above the polar regions.

The event in Canada had been called to discuss a technical solution to solve a measurable atmospheric condition. The chemicals doing the damage were known and the solution lay with an identifiable group: the manufacturers of domestic and industrial freezers and refrigerators. It was relatively easy to prevent any further damage by designing new equipment with different refrigerants which wouldn't migrate to the upper atmosphere and provide for the safe disposal of old refrigerators and freezers and recovery and disposal of their refrigerants.

This is not the case with global warming, since the emissions are much more diffuse. If nothing else there are now three times as many people on Earth as were alive at the end of the Second World War, all breathing in oxygen and exhaling carbon dioxide.

The United States, Europe and Japan, with the greatest concentrations of energy use, are the countries most concerned about global warming and whose emissions have been both increased as a result of industrial development and rising productivity, and been reduced by the improvement of the efficiency of processes and in transport and power generation, and use of nuclear and hydro power, and more recently gas-fired combined cycles, for electricity supply.

Kyoto initially was not a global treaty but one which was directed at what were considered to be the most polluting countries to determine a mechanism of slowing down climate change. The follow-up meeting in Den Haag in 2001 set up a treaty defining what measures should be taken to achieve lasting reductions to 1990 levels. The United States refused to sign because they said that legislation was not the route but technology was. As if to underline this the Finnish Government quietly announced that the only way it could meet its Kyoto targets would be to build another nuclear power plant.

Those countries which ratified the treaty were expected to cut their emissions back to 1990 levels. The target industries were power generation, transport and commercial aviation. In hindsight it is obvious that the Greens played a big role in influencing governments to introduce new green energy measures, and all the while pointing accusing fingers at the United States.

A growing number of wind farms started to appear across Europe and the United States. Some research was started on ways of harnessing tidal energy and wave power from the sea, and solar water heaters were installed in many countries and small photovoltaic installations on larger commercial buildings.

Generating and using energy produces emissions which it would be desirable to reduce; and not just the sulphur and nitrogen oxides which are present with high temperature combustion. But the latest idea is to remove carbon dioxide from the flue gases of coal-fired power plants and the synthetic gas produced for the integrated gasifier combined cycle (IGCC) projects currently under development.

The idea of carbon capture is to stop any more carbon dioxide entering the atmosphere as a combustion product and either store it underground in a deep saline aquifer, or send it to an oil field to be used for enhanced recovery. But if this is the cause of global warming then we are in for a big shock. We cannot apply it fast enough, and if the global population continues to increase there will be many more people breathing in oxygen and exhaling carbon dioxide and driving cars before we have finished.

The question, therefore, is how many coal-fired base-load power stations would have to be fitted with carbon capture units to achieve 1°C drop in global temperature, which is about the order of the global temperature reduction caused by a large volcanic eruption. The chemistry of carbon capture is understood, but we are still a long way from fitting a working system on to a coal-fired power plant which will reduce carbon dioxide emission by even 90%. For a 1300 MW power plant that would be 13 million t/year of carbon. from a reduced net saleable output.

However, all these measures take time to develop and transfer into a commercial system applied to a power plant. There are several new technologies which are in the early stages of development and there are many old power plants which are nearing the end of their service life and will have to be replaced with something.

Governments are beginning to take notice of this and there are concerns about both the future security of electricity supply at present and projected levels. The electricity supply capacity in the 25 countries of the European Union was 704 GW at the end of 2004. Of this 58% was coal-, oil- and gas-fired plant, 19.5 % was nuclear and 18.2 % hydro plant. The rest is wind and other renewables such as solar geothermal and biomass.

This is capacity as distinct from production, because the nuclear plants run at about 90% load factor and the other thermal plants according to need. Two coal-fired plants in Finland, for example, run on a variable capacity factor of less than 20% per year, mainly to cover a shortage of hydro power at times of low rainfall and also outages on the four nuclear plants.

The European Union's Large Combustion Plant Directive is aimed to reduce the acidification of rain, ground level ozone and particulates by controlling emissions of sulphur dioxide, nitrogen oxides and dust from large combustion plant. All plants built after 1987 must comply with the directive if they are to remain in operation after 2015. In the case of older plants which opt out of the directive, they can run a maximum of 20,000 hours between 2007 and the end of 2015.

But the other issue which is beginning to surface again is nuclear energy, with one plant under construction in Europe, in Finland, and a second in France. The industrial power group TVO, has two 730 MW reactors at Okiluoto on the West Coast north of Rauma, which went into service in 1979 and 1981. Their application to build another reactor was voted by the Finnish Parliament in 1994, and it is now under construction alongside the existing units for entry into service in 2012.

Does the fact that there has been hardly any Green protest at any step of this project mean that public opinion has seen through the Green argument and that they are more concerned to have a secure supply of electricity from proven technology? After all, nuclear power plants, because the rection is entirely in the solid state, emit neither carbon dioxide nor sulphur oxides, nor nitrogen dioxide in operation.

From the beginning, the Green movement has associated nuclear energy with the bombing of Hiroshima and Nagasaki more than 60 years ago, yet if any lasting good came out of this it was to create an industry which has monitored the health of its workers and contained its operations in a secure environment to a level that has been achieved by no other industry. Greens argue that nuclear power is inherently unsafe, expensive, and with no proven method of disposing of its wastes. Yet the industry has found answers to all of these problems,

To look at this another way, is the response to global warming an attempt by the same people to force us into a way of life which depends on energy production which cannot be developed in time or cannot produce the energy needed to sustain our standard of living? On the one hand there is the issue of renewable energy, and on the other there is the new technology which will add to the cost of electricity production without improving the performance of the power stations.

So it is still the issue of what is the cause of global warming and how much is it man induced. But that is not to say that everybody has the same level of energy use as those of us in Europe, North America and Japan. A permanent smog which seems to hang over Southeast Asia is as much to do with millions of Indian peasants burning dung and wood (biomass!) for cooking and heating as to thousands of cars clogging the streets of Bangkok and other cities in the region.

While there may be many ways in which individuals can reduce their energy consumption it will be a long, slow process. Individual consumption can be measured in terms of electricity or gas consumed and miles driven. Behind these are all sorts of variables which are by no means predictable: for example a sudden cold snap in spring or autumn could cause heating to be kept on longer. The sudden death of a friend or relative living 300 km away would be an unexpected 600 km round trip by car to the funeral.

Global industry produces some anomalies of its own by transferring manufacture to Asian countries where wage rates are considerably lower. Why, it is asked, do Scottish fishermen send shrimps to Thailand to be shelled, packaged and then returned to Glasgow to be sold in Scottish supermarkets? The blades of the gas turbines powering the aircraft carrying them from Glasgow to Bangkok and back, were probably made in China, again for economic reasons of production. In fact as countries have developed and others have sought to catch up, the principles expounded by the great Conservative philosopher of the 18th century, Adam Smith, has been applied on a global scale.

Smith, in his, *Wealth of Nations*, proposed what we now know as mass production. In his time it was the example of manufacture of a pin. He identified twelve operations. If each job was given to one person in the chain, production of pins would be increased because each job would be supplied to the next person in the chain to do the same operation over and over again and the production of pins would be far greater than if twelve individuals each made one pin in its entirety.

Mass production has gradually moved into automated processes; the car factory, with hundreds of workers feeding components to others fitting them in to the car on an assembly line, is now using robots for many of these jobs: animated computers which can be programmed to perform the same operation with the same accuracy time and time again. Mass production has greatly increased the use of electricity in industry.

Global warming may be inevitable but reduction of emissions is a different matter. Coal-fired power generation has been attacked in Canada and elsewhere on grounds of public health yet electricity as a source of energy is emission free at the point of use. Go to Scandinavia, where in Norway almost all electricity up to now has been generated by hydro power, and to Sweden where it is half water power and half nuclear, and the air is noticeably cleaner. Even in France with a much larger population taking 70% of their electricity supply from nuclear power plants the air, similarly, is cleaner compared with other countries where fossil-fuelled thermal power plants are the main generators.

Transport is the other big emitter. There can be no doubt that the efficiency of the internal combustion engine has improved over the last thirty years and that many people, even in the United States and Canada, are driving smaller cars than they did. But there is an interesting consequence of this in the effect of cars on the rural environment. Drive along a new motorway as part of your regular journey to work and after a few years notice how much trees and hedges planted along the side of the road and down the centre reservation have grown, flooded in the carbon dioxide emitted by the passing vehicles.

So there can be a favorable reaction to carbon dioxide emissions and a case can certainly be made for reforestation, because forests have been cleared for industry, urban development and road building as well as agriculture. There have been instances of power plants being built and an area of trees planted that would take up the equivalent volume of carbon dioxide emitted by the power plant.

Global warming is something which can be sensed, if temperatures are consistently higher than in the past, if deciduous trees come into leaf earlier and shed them later in the year, or if the arctic ice caps shrink as is the case at present. But this comes back to the question as to whether this is a natural tendency because of perturbations of the earth's orbit around the sun or a slight change in the tilt of its axis. Is it a cyclic process with a period of several hundred years? Would we even be talking now about global warming if Krakatau had erupted not in 1883, but in 1983?

The global warming debate has brought plenty of discussion of solutions but little firm action of new plants. On the one hand there are those countries with nuclear power who see the value of it but only Finland and France are building anything. But the European Union's Large Plant Directive, which sets a deadline for action by the end of 2015, is one of the most sensible post-Kyoto measures so far devised.

It may be primarily a measure to promote public health but it depends on what replaces them as to what reduction of emissions results. In France if all the opted-out plants in the country are shut down at the end of 2015, the 1600 MW Flamanville 3 nuclear plant will have been completed and in operation. Elsewhere it is likely to be combined cycles and wind farms that will fill the gap.

If European Governments are serious about cutting emissions, then they have to consider how they can do it. If they keep old coal-fired power plants running to the end of 2015 then that may keep the lights on, but when the plants close, what will be there to supply the energy at the coldest and darkest time of the year?

So what are the options for lower emissions from power generation? There are basically four available: nuclear, supercritical coal, combined cycle and renewables. Upgrading existing plants, which is basically a matter of fitting all or part of the steam turbine with modern computer-designed high efficiency blades, would give about 90 MW more on the output of a 1000 MW set. The same is relevant to any large steam turbine or gas turbine and can be carried out at the regular maintenance outage.

New nuclear plants based on the advanced systems which have already been certificated by European and American governments, means, the advanced pressurized water reactors. Taking the European EPR as the capacity reference, then if all consents are in place the new 1600 MW plant could be up and running in six years on base load with no greenhouse gas emissions whatsoever.

If coal is not immediately taken out of power generation supercritical steam plant is proven technology which if two 800 MW units at 45% efficiency replace a 40-year old 1500 MW plant with three 500 MW sets at 35% efficiency there would be an immediate reduction in emissions of almost 25% with less coal being burnt. If all consents were in place, it would see two new 800 MW generating sets in service in between four and five years.

The combined cycle option is with four 400 MW single-shaft blocks which would use a 280 MW gas turbine and have an efficiency of 58%. Low environmental impact and flexibility of operation can provide the mid-load and peak time power demand to complement the base-loaded nuclear and supercritical coal plants. First unit could be up and running in two years, with the others following at nine-month intervals to give 1600 MW in little more than four years.

Renewables cannot compete. The options are wind, micro hydro, and biomass-fired steam. Micro hydro is limited to there being suitable run of river sites to support a single unit of typically up to 5 MW.

Wind farms are also problematic because the turbines are even smaller and have to be mounted on tall masts and public objection to large wind farms has forced them offshore. There would not be one large wind farm, but probably 16 widely separated with 30 units of the largest currently available, approximately 3.7 MW, in each. This would guarantee that some of them would be running at any one time wherever the wind was blowing. But one cannot guarantee that 1600 MW will be available all year long.

Biomass steam is a practical solution, burning wood chips and other agricultural residues such as rice husks and bagasse, or chicken litter. It is considered carbon neutral because it burns the wastes from this year's crop and the carbon dioxide emitted will feed the growth of next year's crops. Several plants have been with outputs of about 20 MW, which is bigger than that for the other green options. The other source of biomass, is forest and agricultural waste, even pelleted refuse can be mixed with coal and burnt in a power station. It may only account for 4 MW out of 1600 MW, but it will cause the generators to receive the same renewable energy subsidy as the wind farm, or the hydro plant. Most of these biomass coal schemes are associated with the plants that have been opted out of LCPD.

Perhaps we should not be looking at renewables as a generating resource in its own right, but rather as a complementary energy source to the conventional systems. In some cases they would improve the efficiency of the combined system. The biomass option for the coal-fired steam plant is one example. A solar power plant sharing its feed water system with that of a combined cycle is another, the first example of which which is already under construction for a new power plant in Morocco.

The future technology is still under development. Coal, and particularly integrated gasifier combined cycle (IGCC) are inextricably liked with carbon capture, either post-combustion with supercritical steam, or pre-combustion with established chemical processes for the

IGCC. A recent DoE study of IGCC in the United States shows that carbon capture would reduce the output and efficiency of the plant, by up to 6 percentage points and add between 40 and 50 % to the cost per kWh of the electricity produced. There is as yet no IGCC scheme with pre-combustion carbon capture in operation to test these principles, and the first commercial IGCC schemes without carbon capture are not likely to be in service much before 2012.

So these are the Green options: clean coal technology with carbon capture and efficiency lower than that of the 40-year old coal-fired plants that they would be replacing but with only perhaps 10% of the greenhouse gas emissions. Renewables to an impossible level in small units, which cannot achieve the availability of the old coal-fired and nuclear plants that they would replace.

Some of the other proposals of the European Union are equally mind boggling: 26-30% reduction in carbon emissions by 2020 and 60% by 2050. 10% of all transport fuel to come from bio fuel by 2020. These look all the more ridiculous when it is noted that targets for renewable energy by 2010 will not be met and that electricity demand continues to increase. Emission reduction relates to total energy consumption and not just power generation and includes the bio-fuel component of transport.

So if the Green arguments are pushing technology in a particular direction how long will it be before any of it can be applied to have any lasting effect? Or more importantly, have we any guarantee that these new technologies are going to work with the reliability and efficiency expected of them and at an acceptable cost? These are the questions which face utilities in planning for development and the first signs are that this will not be the case.

Since the end of 2002 only one country has specifically said that they must start reducing greenhouse gases now and set about doing it. Finland will have their fifth nuclear power reactor up and generating by the end of 2012. Even now their two coal-fired condensing stations are only used to cover times when, because of low rainfall, there is less hydro potential available. All their other thermal plants are combined heat and power schemes for district heating or industrial combined heat and power schemes.

Other countries whose governments have signed up to Kyoto have gleefully accepted the Green arguments for renewables.

There are signs of electricity supply companies starting to knock down some large old plants to prepare for new ones which at the present time will almost certainly be gas-fired combined cycles. In Europe these are the only plants which can be built in sufficient number and time to meet the 2015 deadline. These will have immediate effect on emission levels since their efficiency is much higher than that of the steam plants they will replace. When they are not generating they can be shut down, so no fuel is burned, and nothing is emitted.

Combined cycle is the only technology which offers a significant improvement in energy efficiency. New gas turbines are under development which will push combined cycle efficiency above 60%, and a greater application to combined heat and power will also improve total energy efficiency. The first combined cycles at the higher efficiency will appear in service before 2012.

There are thus a number of options available with our present technology but there is no evidence that the reduction of emissions that has so far been achieved has had any effect on global warming as the population has continued to grow.

But there are other measures that can be taken individually by consumers everywhere. Building codes in the UK have been re written to cut energy demand for heating. Wall and roof insulation and double glazing are now standard on any new house. Low energy lighting is another benefit which can reduce household electricity demand, and all can be retrofitted on existing buildings.

But the other issue of green technology is the energy use in setting it up. because it requires large numbers of low output devices. A recent British Government proposal to plant seven thousand of the largest available wind generators, 3.7 MW, in offshore wind farms around the coast to supply every home in the country with electricity from renewable power sources by 2020 is highly dubious, given the logistics involved of building the components and assembling them out at sea at the rate of two units a day and then connecting them up to the grid.

Wind farms have no greenhouse gas emissions but there is an lot of steel in 7000 units which accounts for a huge expenditure of energy to produce the materials, assemble the components and install them out at sea. With a total capacity of 24,500 MW between them these wind generators would on an average availability per year of at best about 30%, would produce 6.4 TWh per year.

At the end of 2007 the US Government passed a bill to take incandescent light bulbs off the market. This immediately produced reactions. On the one side were the Human Rights Activists who said that the measure was a fundamental denial of a human right to determine what sort of lighting to install in their house and garden. A more measured response was that if enough lights were changed the reduction in demand would be equivalent to the output of four large coal-fired power stations. If one million units were changed in each of the 50 states the lighting load over the whole country would drop by about 4000 MW.

This is a practical example of technology at work. Fluorescent lighting of public buildings and advertising displays have been around for more than fifty years. It was only a matter of time before it entered the domestic market. The long tubular fluorescent lights were already being installed in kitchens and utility rooms before the low energy "bulbs" came on the scene with a single bayonet type or Edison screw fitting so that they could be put into a standard ceiling fitting in any room of the house.

So to phase out incandescent fittings is going to have a significant effect on demand for electricity for lighting, particularly in northern latitudes. Quite simple technology applied on a large scale can make a significant difference to the rate of growth of electricity demand so that new construction is primarily to replace old plants with new having higher efficiency and fewer or no emissions.

3

Carbon capture and storage

Carbon capture and storage (CCS) is an established practice in the oil and gas industry with a number of projects using carbon dioxide for enhanced oil recovery. Separation of carbon dioxide is a widely used chemical process. In each application the gas has a commercial value, but as the separated waste product of power plant flue gases a much larger volume of gas might have no value at all.

Herein lies the problem. CCS is seen as the means to remove carbon dioxide from the flue gases of power plants and particularly coal-fired plants. But it will take several years to bring the technology up to the stage where for a power plant with two 800 MW coal-fired supercritical steam sets it is capable of separating out at least 90% of the carbon dioxide emissions.

Given that the life of a coal-fired power station is about 40 years, by the time the full scale technology has been applied to an existing coal-fired power plant, and proven to work, assuming it to be in about 2020, the only plants that could justify having it would be base-loaded power plants built after 2000.

Given the low efficiency of the modified power plant which results from the heavy auxiliary load of the CCS equipment, it might only be one large supercritical steam plant that will be so modified to test the principle.

Until now the improvements to coal-fired power plants in service have been designed to achieve higher performance and remove harmful trace elements from the flue gases. Low NOx burners, FGD and electrostatic precipitators clean up the exhaust, and the refurbishment of the steam turbine with modern computer designed blading and new bearings raises the output. The higher pressures and temperatures of the supercritical steam cycle raise output and efficiency to compromise for the loss of performance resulting from the fitting of FGD which however produces a saleable commodity, gypsum, to help offset its costs.

Turbine rebuild is also associated with repowering, since the condition of the boiler depends on the way the plant is used. Most coal-fired plants are run on base load in their early years, and as new plants, which are more efficient and have lower operating costs, come into operation, the older units are pushed down the merit order and are used more for load following and frequency control. All the while it has been operating, from day one the plant has started to wear out, and this speeds up with more varied modes of operation. There will come a time when the running hours are low and the efficiency has declined.

To replace the original fired boiler would not greatly improve the performance but to repower the plant by replacing the old boiler with a gas turbine and heat recovery boilers would have an immediate benefit in higher efficiency, lower emissions and move the plant back up the merit order where it could earn more for higher output over longer running hours as a combined cycle.

Those who advocate CCS are mainly looking at coal fired capacity, believe strongly that coal is the fuel of the future, and are aiming to use the new clean coal technology to stabilize carbon dioxide in the atmosphere.

The fact that the oil and gas industries are in the forefront of CCS development is because carbon dioxide is, to them, an impurity which they must remove from oil and natural gas to meet their customers' specifications. When natural gas has to be liquefied for shipment to market it is cooled down to a lower temperature than the sublimation point of carbon dioxide which, if it was not first to be removed from the gas, would cause blockages in pipes and valves of the liquefaction plant.

So there are good commercial reasons for removing carbon dioxide and the oil and gas fields have the geological structure to contain the sequestered gas and prevent it leaking up to the surface. Among projects currently in operation, that at In Salah, Algeria, was developed as a joint venture by BP, StatOil and Sonatrach in response to Kyoto.

BP's chief executive at the time of Kyoto at the end of 1997, Lord Brown was one of the first to see the connexion between global warming and fossil fuels, and decided to cut annual emissions from BP operations by 10% from a 1990 base line. For a company like BP it was relatively easy to make the reduction because they were already stripping out carbon dioxide from natural gas and the Algerian project offered the best opportunity, because it would completely stop the emission of the separated gas to atmosphere.

Current gas production in Algeria is from three fields: Tegentour, Reg and Krechba, with reserves of 16 billion m³. Krechba is the northernmost of the three producing fields and is the site of the processing facility. The gas contains up to 10 % carbon dioxide which is stripped out with a reusable amine solution to meet a European sales specification of less than 0.3%.

The gas zone at Krechba is about 20 m thick in a layer of carboniferous mudstones above a deep saline aquifer about 2000 m below ground. Carbon dioxide from the three fields is separated and sent to three injection wells which pump it at 185 bars back into the aquifer at its lowest point. The same impermeable rock structure above the aquifer stops the carbon dioxide escaping and returning to the surface just as it capped the gas deposit before it was discovered.

At present rates of production from the three fields, about 1 million t/year of carbon dioxide is being stored at In Salah which is a small quantity compared with the estaimated 27 billion t/year/of greenhouse gas emissions in 2007.

StatOil, who were BP's partner at In Salah have instituted their own CCS project on their Sleipner field in the North Sea, and have completed another at their Arctic Snøhvit field which came into production at the end of 2007. Norway in 1991 introduced a carbon dioxide offshore tax to curb emissions. Sleipner West came into production in 1990 with the gas containing 9% of carbon dioxide. So it was a tax issue which drove StatOil to remove and store the gas. The equipment was too large to mount on an offshore platform but it would remove 1 million t/year of carbon dioxide which was then equivalent to 3% of the total Norwegian emissions. It would mean building the separation plant on shore and installing a pipeline and compressors to take the gas away.

The process uses the monoethanolamine (MEA) solvent whereby the amine is combined with the gas which is then boiled off leaving the amine absorbant to be reused. The Sleipner platform is above Utsira, a deep geological formation consisting of a very porous sandstone filled with salt water 1000 metres below the sea floor. Since the scheme began in 1996, some 9 million tons of carbon dioxide have been stored there.

The Snøhvit solution is similar except that in this case almost the entire natural gas production is liquefied for shipment, and so carbon dioxide has first to be removed. As at Sleipner the MEA process is used to recover about 700 000 t/year of carbon dioxide which is sent 145 km back to the field where it is injected into the geological formation called Tubåen some 2500 m below the sea floor and well below the gas reservoir.

A third Statoil project addresses the requirements of the electricity market. The Mongstadt refinery near Bergen is to be the site of a new gas-fired combined heat and power plant which will supply 350 MJ/s of heat to the refinery and so allow much of its existing boiler and generating plant to be shut down. Power will be supplied to the refinery, the Troll A and Gjøa platforms in the North Sea, and the Troll processing terminal at Kollsnes.

The power plant, which is owned and will be operated by Dong Energy, of Denmark, will burn Troll gas, and will come into operation in 2010. When this happens a CCS process with a capacity of 100 000 t/year will also come into operation. A full scale carbon capture plant is is due to come on stream in 2014 and to serve both the power plant and process sources in the refinery.

Norway by taxing the offshore producers has given a kick start to carbon capture and storage. Not only that, but by providing more efficient generation plant to take over it will achieve an even bigger reduction in emissions.

In Canada, the EnCana Corporation's Weyburn field in southeastern Saskatchewan is the first instance of the international sale of carbon dioxide for enhanced oil recovery from a third party. The Dakota Gasification Company, who own the Great Plains Synthetic Fuels plant in Beulah, ND, supplies 6000 t/day of carbon dioxide which has been separated from the coal gasifier output and piped 325 km north over the border to Weyburn.

The field which is some 16 km southwest of the town of Weyburn, has been in production since 1954, and by 2000 the daily production rate was 18000 bbl/day which was assisted in later years by water flooding. By then some 340 million bbl of oil had been recovered. Carbon dioxide was introduced the same year at an initial rate of 2.69 million m³/day to continue oil recovery.

Since then the rate of injection of gas, which includes that recovered from oil production, has increased and by the end of 2008 some 10.8 billion m³ will have been stored there. EnCana saw enhanced oil recovery with carbon dioxide giving the field another 20 years of life and a further production of 130 million bbl of oil.

Being on land in North America the field became the subject of an extensive study under the IEA Greenhouse Gas Research and Development Programme, based at Cheltenham, in the UK. At Weyburn the study was of the behavior of carbon dioxide storage in relation to enhanced oil recovery.

The study covered an area of 400 km² and sought to determine how much of the injected gas went to oil recovery and how much to storage and how much it might migrate to other strata, projected over a time scale of 5000 years. Results show that there is capacity for approximately 45.1 million t of which about 20 million t will have been taken up with enhanced oil recovery, and that there is no likelihood of any carbon dioxide escape into the biosphere.

These oil industry schemes are the only examples of carbon capture and storage and are on a relatively small flow rate. Since the gas has to be removed anyway, and there is use for it in enhanced oil recovery, it has a commercial value.

In Texas where carbon dioxide has been found in naturally occuring deposits it has been extracted and used, together with some industrial output, for enhanced oil recovery and the operators have benefited from a tax break, as a result of which some 33 million t of carbon dioxide have been stored in Texas and New Mexico over the last 40 years.

These are all commercial schemes encouraged by tax concessions or with a commercial value for the carbon dioxide. The economics in some cases may be marginal but they offer significant environmental benefits and this is what is driving the technology forward.

Stripping carbon dioxide out of power station flue gas is totally different and on a much larger scale. A coal-fired supercritical steam plant of 1600 MW would produce up to 16 million t/year of carbon dioxide and may not be conveniently close to an oil field. In this situation the recovered carbon dioxide would therefore have no commercial value at all.

Even if there was a conveniently sited chemical works or a fertiliser production plant in the neighbourhood, the amount of carbon dioxide produced in a year would be far less than that from the power station. What then if there is more than one coal-fired power station on a network serving an area where there are no oil fields, and an insufficient industrial capacity to absorb all the carbon dioxide produced?

Already coal-fired stations are coming under attack by Green activists, the same organizations that thirty years ago didn't want nuclear energy, no doubt because the coal-fired plants cannot be built with CCS and because no proven system exists on the scale required. It means nothing to them that the new power plants are much more efficient than those that they would replace and would produce the same quantity of electricity with lower emissions.

Attempts to cut carbon emissions have been economic rather than technical. Some countries have introduced carbon taxes to encourage improvement in energy efficiency. In the United Kingdom the cost of the annual Road Fund licence for cars is determined by the exhaust emission rating in g/km of carbon dioxide.

The other measure is termed a Cap and Trade system which is a system of permits which are traded from non-polluting industries to those with high emission levels. So if the electricity supply has a large component of nuclear and hydro plant as for example in France, Belgium, and Sweden the operators gain from the sale of permits to the coal and gas-fired plants to keep them running.

If we are serious about cutting carbon dioxide emissions we do it by moving to more efficient generating systems with much lower or no emissions. The European Union Large Combustion Plant Directive which comes into operation at the end of 2015, has already led to the building of combined cycles to replace the coal-fired plants that must be shut down and which in any case have been limited in their operating hours since 2007.

None of these new plants will have CCS fitted but will have in any case much lower emissions. Also given the huge number of sites that would have to be supplied with CCS could the job be done quick enough to see any reduction in climate change?

The answer must surely be no, because of the time that must be taken to prove the technical and economic worth of a CCS system. This is quite a large installation, and the prototype system at Pleasant Prairie, WI, which started operation in February 2008 is only taking a slipstream off one flue to test the principle.

CCS is in fact a process plant stuck on the back of a power station. But it also has to dispose of the captured carbon dioxide which requires compressors and a pipeline to a distant oil field or a suitable geological site. All this equipment has to be produced at great cost in energy and if the captured carbon has no commercial value, as would be the case if it were stored in deep underground saline aquifers, the cost of that equipment along with the extra cost of fuel will have to be passed on to the consumer.

If we concentrate on power generation then the immediate need is for an effective post-combustion system for the coal-fired power plants that provide much of the electricity supply at present. But many of these will have been taken out of service by the time that the technology is proven both practically and economically which may not be for another ten years at the minimum.

The large supercritical steam plants built in the last twenty years must be the priority because they have the longest remaining life of the current coal-fired plants. By 2015 these may be the only coal-fired plants remaining in operation. In Europe, at least, most of the coal-fired plants built before 1987 will be shut down under the provisions of the

Large Combustion Plant Directive. Those that survive will have had FGD fitted and other environmental measures installed before the end of 2007, but of those that have opted out some could be used for testing CCS systems for the few remaining years in operation.

In studying the application to existing power plants the important factors are the resulting change in performance and the additional costs. FGD was less of a problem in this respect because the end product of the process was gypsum for which there is ready market in the building industry to make plaster board. But with CCS, unless there is a market for the separated gas, it has no commercial value. Whether it is sold to an oil field or simply consigned to a deep underground saline aquifer, the gas still has to be moved, which means that pipelines must be constructed and boreholes and compressors to drive the gas to storage.

All studies of CSS to date have revealed large auxiliary loads and significantly lower outputs of saleable electricity produced at lower efficiency. Besides the electric motors driving the system the auxiliary loads are primarily the compressors for the separated carbon dioxide and the steam demand from the turbine for the boiling off of the gas from the absorber.

Most studies in the United States have used the MEA monoethanolamine solvent technology but American Electric Power, one of the country's largest coal-burning utilities has thrown it weight behind the Chilled Ammonia system which would appear to offer less reduction of output and efficiency.

Whichever separation technique is used, the loss of output and the need to make it up some other way could add as much as 50% to the price of electricity. This is an unacceptable price which would be applied by stealth. Consumers would not notice it until a sufficiently large number of modified plants were in service and accounting for a large enough share of the electricity supply. The first, and maybe the only ones, to be converted would be the large supercritical steam plants supplying base load.

The other issue of CCS is that it is not confined to coal-fired power generation but to combined cycles, gas turbines and industrial boilers. Recent developments in Europe and Japan have looked at gas-fired power plants which have much lower emissions. Applied to gas-fired industrial combined heat and power schemes, there could in some applications be a use for the recovered gas as a chemical feedstock

Furthermore it is under development at a time of increasing financial turbulence with a new threat of inflation which has largely been kept under control for the last fifteen years. What threatened inflation was

TABLE 3.1: EFFECTS OF CCS ON PLANT OUTPUTS

	Without CCS	MEA System	Chilled Ammonia
Coal feed rate, t/h	151.3	151.3	151.3
Heating value, kJ/kg (HHV)	23491	23491	23491
Boiler heat input GJ	4200	4200	4200
LP steam extraction kg/s	0	153.2	22.6
Gross plant output MW	491.1	402.2	471.3
Plant auxiliary load MW	29.0	72.7	53.9
Net power output MW	462.1	329.5	421.7
Net efficiency HHV	40.5	28.9	37.0

oil at over $140/bbl, wheat at over £100/ton from a harvest reduced by cultivation of bio-fuel crops, and a general increase in commodity prices due to increased demand from the rapidly developing economies of China and India.

Power generation with fossil fuels accounts globally for 40% of greenhouse gas emissions from the energy sector followed by transport with 24% and industry with 22% . In power generation 70% of the emissions are from coal-fired plants. Therefore to work on a scheme to phase out coal from power generation, and replace it with nuclear gas and hydro, supported by other renewables would perhaps produce a much greater reduction of emissions than CCS on a few base-loaded power plants installed half way through their working life.

But for the general public to accept inflation in the cause of "saving the planet" it must be a visible change for the better. But given the nuclear experience where 450 power plants with no emissions at all along with the millions of tons of carbon dioxide that have been used for enhanced oil recovery over the last thirty years, have made no difference to the rate of increase of global temperature during that time, we can expect CCS to show the same effect particularly if it cannot remove all of the carbon dioxide. But as the table clearly shows, CCS is the first environmental enhancement which drastically reduces the output and efficiency of fossil-fired power generation.

Table 3.1 shows the effect of a full scale CCS scheme on the performance of a nominally 500 MW supercritical steam set in the United States, burning the high-sulphur Illinois 6 coal. Without CCS the net output is 462 MW at an efficiency of 40.5%. With the MEA system, which is used in the oil and gas industries to remove entrained carbon dioxide for enhanced recovery, the output drops to 329.5 MW at 28.9% efficiency. The chilled ammonia system operates at a lower temperature with a smaller process energy load. Even so the saleable output is 10 % lower but the efficiency at 37% is only marginally higher than that of a subcritical steam station of the same nominal output.

Efficiency is measured as the ratio of the net electrical energy output to the fuel energy input. What it means is that for the steam plant without CCS, for every ton of coal converted to electricity, the energy of 1.47 tons are thrown away in the cooling water and the flue gases. For the MEA equipped plant the energy of 1.66 tons is lost to produce 143 MW less of saleable electricity.

Of the various systems that have been tested the most promising appears to be the Alstom chilled ammonia system which is a low temperature, high pressure process that has been installed for an American trial in conjunction with the Electric Power Research Institute (EPRI) at We Energy's Pleasant Prairie power plant.

Pleasant Prairie is a 1224 MW subcritical coal-fired power station located near the village of that name about midway between Chicago and Milwaukee and some 8 km back from the Lake Michigan shore in Wisconsin. The station runs base load and burns low-sulphur western coal. The two tandem compound reheat turbogenerator sets are each rated at 612 MW with steam conditions at the high pressure stop valve of 137 bars 535°C. The first set entered service in 1980 and the second in 1985. For the CCS study a slip stream is taken off one of the flues equivalent to 20 MW of generation.

The basic CCS system is the same however it is applied, the difference being in the solvent to separate the carbon dioxide and the quality of the flue gas. The incoming flue gas has first to be cleaned and cooled to less than 30°C a water wash removes sulphur and nitrogen oxides and any particulate matter and cools the gas to the optimum temperature for capture by the absobent.

The clean gas then flows up a column in counter current to the descending slurry containing the absorbent which separates the carbon dioxide leaving the rest of the flue gas to go to the stack. The enriched solvent then passes through a heat exchanger to cool the lean solvent leaving the separator.

The separator heats the enriched solvent causing it to give up the carbon dioxide, becoming lean to return to the top of the absorbent column. The separated gas passes through a water wash to remove any of the solvent that may have carried over and is 99.9% pure. It is then compressed and continues to process or sequestration. The energy demands of the process are in heating the absorbent to release the carbon dioxide and the compression of the separated gas.

Most of the processes use an organic solvent, of which the most widely used is monoethanolamine. But Alstom's chilled ammonia system was designed from the outset for power generation and is a high pressure, low temperature system, which uses ammonium carbonate as the absorbent.

In the first stage, the incoming flue gas from the FGD unit is at typically about 60°C and is cooled down to about 2°C which condenses out most of the entrained water which can be recycled back to the FGD system. This reduces the volume and mass of the flue gases and reduces the size of the following equipment and of course the compression power required.

The clean gas with about 1% moisture enters the absorber column which is similar in concept to the FGD unit. The gas rises in counter current to the slurry containing ammonium carbonate and ammonium bicarbonate, which absorbs more than 90% of the carbon dioxide. The de-carbonized flue gas passes through a cold water wash which removes any traces of ammonia which are recycled.

The carbon dioxide-rich slurry consists mainly of ammonium bicarbonate which is pumped to the regenerator, through a heat exchanger which heats it to 80°C to release the gas and leave a lean ammonium carbonate mixture which is sent back to the absorber through the other side of the heat exchanger.

Finally the recovered gas is passed through a further water wash to remove any traces of ammonia and is then compressed to between 80 and 100 bars for transmission to its final destination. The stack emissions are mainly nitrogen, excess oxygen and traces of carbon dioxide. For a new plant the stack would probably be incorporated in the axis of the cooling tower to give the gas flow extra buoyancy in the vapour stream coming off the tower.

The advantage of the chilled ammonia system is that the lower volume of dry gas passing through it reduces the compression power and the steam requirement for heating to boil off the carbon dioxide in the regenerator. This results in a lower auxiliary load that increases the electrical output of the steam turbine. In fact the steam bleed from the turbine is only 22.6 kg/s about one seventh of the requirement of the MEA system.

The pilot plant has been in operation at Pleasant Prairie since early February 2008 and will continue for about a year. It was designed, built, and is operated by Alstom. The purpose is to demonstrate the system operating on flue gas and collect data which would enable a full size CCS system to be designed for commercial application.

3.1: Pleasant Prairie, WI, USA. The 5 MW CCS pilot plant for the Alstom chilled ammonia system which went on test at the end of February 2008. (Photo courtesy of We Energy)

First reports from Pleasant Prairie show that the chilled ammonia can separate between 88 and 90% of the carbon dioxide and achieve 99% purity, This is one of three similar schemes with slipstreams on large coal-fired power plants.

American Electric Power with 38,000 MW of mainly coal-fired capacity in eleven States is one of the largest electric utilities in the United States. The company has been particularly active in the environmental aspects of power generation. In 1990 it installed the PFBC gas turbine at their Tidd plant near Cincinatti, OH, and in 2002 they signed an agreement with Batelle to study CCS for their 1300 MW Mountaineer power station in Newhaven, WV.

In particular the geology of the area was to be studied to see if it had a suitable deep lying strata for carbon sequestration and also the integrity of the capping rock structure to see if it was in any way fractured which might cause the gas to leak up to the surface.

Then in 2007 AEP signed a memorandum of understanding with Alstom for the installation of a Chilled Ammonia system at Mountaineer. Like Pleasant Prairie it would be a slipstream taking flue gas equivalent to 20 MW of generation. The recovered gas, at about 200 000 t/year will be piped to a deep lying saline aquifer beneath the site. A system for the full output of the station would have to handle 13 million t/year of gas.

A third project is for a 450 MW unit of the Northeastern station at

Oologah, OK, which will be a similar size to the unit at Mountaineer. Here, AEP have signed an agreement with SemGreen LP who will take the carbon dioxide for enhanced oil recovery.

The use of slipstreams to take say 5 or 10% of the flue gases for CCS rather than100% may have something to do with the fact that the Chicago Carbon Exchange is the first and the most effective carbon trading scheme to be set up post Kyoto. Under it, non-polluters sell their pollution permits to polluters who cannot quickly shut down their operations. So a 600 MW set with a 10% take off of flue gas to pass through a CCS system has effectively 540 MW of classic coal-fired dirty power, and 60 MW of clean power. Split the output into these two parts and the 60 MW sells their pollution permit to the the other 540 MW which is purely an internal transaction of the power plant.

But CCS is approaching a crossroads. The Pleasant Prairie and other tests are aimed for EPRI to produce a report on the suitability of the scheme for scaling up to a full size power plant. What would be the operation and maintenance issues that could be expected, and above all what would be the cost, because this is ultimately what would determine whether it would be built as a full size carbon capture scheme for a supercritical steam plant.

But the other problem which is still to be addressed is what is the auxiliary load on the power plant installing it. Say, for example, that a company has on its network three large supercritical steam plants, each with two 800 MW sets at an efficiency of 43%. They are obliged to install CCS systems to remove the carbon dioxide from the flue gases of all six units. These are base-load plants so are producing a total of about 48 million t/year which has to be stored either in a nearby oil field for enhanced recovery, or in a deep-lying saline aquifer which could be anywhere.

The equipment is installed over a 3-year time scale and the net result is a saleable output of 700 MW from each set which is a net loss of 600 MW over the three stations and the efficiency has dropped from 43 to 36%. But before conversion, all three power plants were running as base load units, and each had contracts to supply 800 MW. So how does the utility make up the difference? It could buy in the power from a neighboring system or it could build an 800 MW combined cycle plant to top up their base load capacity and allow for some growth.

The converted coal stations have been cleaned of their emissions but to do that they produce less saleable electricity with the same amount of coal. More energy must be spent in pumping the recovered gas to a distant oil field or a saline aquifer 2000 m or more below the ground.

The big problem is cost and feasibility on a given site. To repower a steam plant with a gas turbine the use of a vertical heat recovery boiler has often been neccessary because it has a smaller footprint and can fit into the space vacated by the old fired boiler. Now with a coal fired plant the CCS equipment has to be installed on the site, but where, given that it is included in a process chain with the electrostatic precipitators and the FGD system. Then, inside the power plant, the steam turbine has to be modified to provide a steam take-off for the absorbent recovery. Given the cost of all this and the large loss of saleable energy which cannot be offset against it, and the age of the plant which will determine its operating pattern, it is likely that the owner of an old coal-fired power plant on a low load factor would prefer to shut it down and put a combined cycle in its place.

The problems would appear to be worse with IGCC, where the syngas from the gasifier is mainly hydrogen and carbon monoxide. By using a high temperature shift reaction of the carbon monoxide in the syngas with steam it produces hydrogen and carbon dioxide. It is proven technology in that it is used in the Ammonia industry, but on a much smaller scale, and to integrate it into a 700 MW IGCC scheme it would be a continuous process to separate a much larger volume of gas. In this situation the synthetic gas reaching the gas turbine is therefore hydrogen with traces of methane and carbon dioxide.

The combustion system of the gas turbine will need to be modified to burn hydrogen, but it will have still to be started with natural gas or distillate and then changed over. The gas turbine output would be less on hydrogen, but the combustion product is water leaving the stack as billowing clouds of steam.

At the end of 2006 the US Department of Energy produced a report which showed that a 640 MW IGCC with carbon capture would have a net output reduced by an additional 100 MW and would add at least 36% to the price of electricity, but yet would produce the cheapest power compared with coal-fired supercritical steam plant and combined cycle each with CCS added.

Without carbon capture the auxiliary load of the gasifier is about 130 MW for a plant with a gross output of 770 MW. Apply carbon capture and it drops to 745 MW because the gas turbines have a lower output when burning only hydrogen. But the additional load of the carbon capture equipment is 60 MW and the net output of the combined cycle is therefore 555 MW. More important is the fact that the net efficiency drops to 32.5%.

At the end of 2007 DoE pulled out of funding Futuregen. This was an international project to build an emission-free coal-fired power station. It was to be built on a site at Mattoon, IL which had been announced about two months earlier. DoE had spent more than $50 million on various studies and had come to the conclusion that it would be cheaper to fit CCS to existing power plants for evaluation rather than build a dedicated facility at a projected cost of $1.5 billion.

Futuregen would not have been operational until about 2012, at the earliest and project definition had not got as far as choosing the suppliers of the major equipment. Meanwhile development was going ahead of CCS pilot plants, of which the first was about to start operating at a coal-fired power plant in Wisconsin.

What is more to the point, who is insisting that we take all the carbon dioxide out of the flue gases of every thermal power plant when, for base load power, there is a proven system of generation with forty years operating experience, lower fuel costs, and no emissions of any sort? If this is the best that we can expect from IGCC with carbon capture it negates everything that has been done to improve the efficiency and cleanliness of emissions of coal-fired power plant for the last fifty years. Throughout history the purpose of development has been to improve performance and efficiency. Yet to strip out carbon dioxide requires the development and construction of a system which can be integrated with an existing power plant at some indeterminate future date.

But what if your power system has a large nuclear base load supported by combined cycles as, for example in Japan and Korea? For the past sixteen years Mitsubishi Heavy Industries in association with the utility Kansai Electric has developed the KM CDR (carbon dioxide recovery) process for flue gas streams from natural-gas-fired boilers, together with the KS-1 absorber. KS-1 has been tested for over 4000 hours on a coal-fired boiler using a 10 t/day slip stream.

The other amine solvent in common use, MEA, tends to degrade relatively quickly, whereas KS-1 a sterically inhibited amine, has proved to be more stable although impurities in the gas stream can react with it to form salts which do not break down under heating and in time reduce the amount available for gas absorption.

The first commercial installation was at the Petronas Fertilizer plant at Kedah, Malaysia. To make fertilizer from natural gas requires the production of ammonia followed by reaction with carbon dioxide to produce urea. The ammonia and carbon dioxide are produced in a steam reformer with the ammonia excess to the needs of the reaction. If the boiler producing the steam is fitted with a CDR scheme the recovered carbon dioxide can be used with the excess ammonia to make more urea.

The boiler produces 3400 Nm³/h of gas which leaves through the CDR as a gas of 99.9% purity.

This is a case where the recovered gas is used directly in the process to improve yield. The plant at Kedah went into operation in October 1999 and removes 200 t/d of carbon dioxide from flue gas to use in an industrial process. Degradation of the KS-1 solvent is about 300g/ton of gas and attempts are being made to reduce it further. Following Kedah a 330 t/d installation was made on a dual-fired boiler at a chemical plant in Japan. When oil is being fired there can be sulphur and and other impurities which must be washed out of the flue gas before it can be treated. KS-1 solvent is used and the recovered carbon dioxide is used in various processes.

Then two 450 t/d units were supplied to Indian Fertilizers at Aonla. It was a similar but larger installation than at Kedah making urea for fertilizer from natural gas. The CDR units went into operation in December 2005. Abu Dhabi Fertilizer also have a 400 t/d unit which went into operation in October 2006.

These units are all operating and have proven the concept of carbon dioxide recovery as a feedstock for urea. These are relatively small units of less than 1 million t/year of recovered gas, but they show that any industry with a requirement for carbon dioxide as a feedstock can recover it from their own gas-fired boiler plant.

Also development has been undertaken by a company making gas turbines acting in partnership with their customer, an electricity generator. They had a vested interest in developing a system which was suitable for gas-fired power plants and which would not impose a large auxiliary load. But it has yet to be tested at full scale on a combined cycle.

Combined cycle has two advantages for Carbon Capture. Its levels of emission are lower and natural gas is a clean fuel. NOx can be reduced in the gas turbine combustion process, with further selective catalytic reduction, to 3 vppm. Second, clean fuel and low environmental impact mean that it can be installed closer to towns and industrial sites where there could be a demand for carbon dioxide for use as a chemical feedstock. But although a current F-class gas turbine in a 410 MW single-shaft combined cycle has a carbon dioxide output from natural gas of 3100 t/day, it compares with 450 t/day for the fertilizer plants. So again the industrial plant is much smaller.

Here we see the disadvantage of Carbon capture. If it is fitted to an existing plant it will cause a loss of saleable output which cannot be recovered. Therefore the process has got to be developed to the point that the output loss is much lower, and that the degradation of the absorbent must be reduced.to lower the operating cost.

But by the time a system has been perfected for the much greater gas outputs of power generation and can produce competitive power, and it could be another twenty years, how many of the existing coal-fired power plants will have been closed by that time?

All the advantages can be applied to the existing applications and the question now is should it be these industrial boilers, and the oil and gas industries, on which we should be concentrating, to improve the efficiency of their existing systems? Almost certainly it will not be applied to the full output of an existing power plant because it would so modify the performance and the resulting price of electricity that probably no generating company would want to install it.

A 1600 MW coal-fired station with two 800 MW sets would be producing 16 million t/year of carbon dioxide. A slipstream from one equivalent to 5% of the output would have a CCS system producing about 400 000 t/year which could be sold to a local industry or an oil field would be in the same situation as another plant taking 5% of its fuel input as biomass.

Biomass being carbon neutral the power plant would certainly in the UK, get a premium price for renewable energy. Would the CCS equipped power get a premium price for its 5% of cleaned coal energy. After all, if by fitting the CCS equipment the plant emissions are reduced by a further 5% that is surely worth something if only as a zero carbon trading permit to offset against the carbon payment for the other 95%.

But if in order to make coal-fired power acceptable the cost is a 50% increase in the price per kilowatt-hour, are the generating companies going to accept it when CCS presents a huge auxiliary load which will require a new plant to be built to make up for the lost output?

Even now with the fuel price rises in place, electricity rates have been put up in many countries and after a long period of low inflation people are starting to see it rising again. They are not going to take kindly to a government which spends millions of dollars to develop a system which will take all the carbon dioxide out of flue gases and then spend as much again on pipelines and compressors to send it to some subterranean saline aquifer which has a finite capacity to accommodate it.

Coal-fired power plants produce too much carbon dioxide for disposal. If a 1600 MW coal-fired plant were to be built in the vicinity of Weyburn, Saskatchewn, they would be wanting to put 16 million t/year into the oil field which would have room for only another 25 million t after the field closes in 2020. Even if no gas had been introduced for oil recovery, the field would have been able to accept 45 million t which would be less than three years output from the power station.

So while saline aquifers can be used and the long term reaction under pressure of the carbon dioxide is understood, how many are there in the geology of the coal-burning countries and have they the capacity to take all the separated gas from all the coal-fired power stations for all of their operating life? This something we do not know but if there is not enough capacity in potential sites for sequestration then this another obstacle against clean coal.

One must therefore ask what is CCS about? Is this another example of Green fanatics realising that there is an established process working on a small scale in the oil and gas industries which is just what we need to stop carbon dioxide emissions; and not just from coal-fired power plants but combined cycles and industrial boilers burning coal or gas?

CCS must therefore, if it is used at all, be used on industrial boilers and combined heat and power schemes burning gas where there is a market available for the carbon dioxide as a chemical radical. There is every reason to continue research into the system to develop higher pressure processes, and more stable adsorbants. But this is just the normal development that one would expect to improve performance and reduce operating costs.

Combined cycles have much lower emission levels; first because of the different fuel; second their high efficiency meaning that they burn less of it; and finally that as load followers in the future green energy scheme they will not be running 24 hours a day on 365 days a year.

Therefore stripping carbon dioxide out of exhaust gases is not sensible because of the huge volumes to be sequestered, and the loss of performance and greater fuel consumption needed to sustain existing levels of electricity demand.

Whether or not to put CCS on existing plants depends how they are used. Centrica Energy, of Windsor UK has seven combined cycle plants and an eighth under construction near Plymouth but they operate them in load following mode to meet the demands of their domestic market to which they also supply gas.

The Centica operation is typical of the probable way that combined cycles will be used in a green energy system of tomorrow. In 2006 seven power plants totaling 3269 MW of combined cycle capacity emitted 4,341,366 t of carbon dioxide which is about the same as would be emitted by a single 400 MW coal-fired generating set which would be running base load.

So to put CCS on a combined cycle is impractical for two reasons.

Firstly, a single shaft block with an F-class gas turbine and an availability of 33% would be running 11 hours a day on weekdays and shutting down at weekends. There is no evidence yet as to how a CCS unit would perform in an intermittent fashion, and how the combined cycle performance would be affected.

So to reduce carbon dioxide emissions from flue gases there are only three practical measures. First, all carbon dioxide required as a chemical reagent in industry should be recovered from the industry's own boilers and power plants. Second, increase the efficiency of existing fossil-fired power plants either by progressive upgrades of existing gas turbines and steam turbines or design new combined cycles for higher efficiency. Finally, take coal out of power generation altogether.

If we are serious about reducing emissions then these are the easiest measures to take. There are too many unknowns about carbon capture and storage and at the present state of the art, it reduces the efficiency and the net power output by so much as to make it unattractive to the generating companies.

Why should we ever be attempting this exercise to render coal environmentally friendly when we have another base-load emission-free generating system which is proven in operation for more than forty years?

4
The end of coal?

Of all the known fuels, coal has occupied a special place in the countries that produce it. It made possible the industrial revolution as a fuel to produce steam to drive pumps, trains, ships, power plants and for domestic and industrial boilers. It is the most abundant of the fossil fuels, but the most difficult to extract, handle and store. It has killed many of the people who mined it and destroyed the health of most of the rest, and burning it has brought smog to the cities, and acid rain to lakes and rivers.

In the last fifty years it has lost most of its traditional markets, leaving only power generation and parts of the steel and chemical industries as its principal uses. Now its days are numbered in the power generating market because all the effort to clean up its emissions, starting with flue gas desulphurization, has reduced the performance and increased the cost of construction and therefore of the electricity produced.

Yet it may have a future market as a chemical feedstock when oil runs out or becomes too expensive. But if any government is serious about cutting emissions to thwart climate change, then the removal of coal as a fuel for power generation must have a high priority.

For a long time coal was the main fuel in the majority of homes in the United Kingdom. A typical three-bedroom house built in the 1930's would have fireplaces in the two main living rooms on the ground floor and in the two largest bedrooms above them on the upper floor. There would also be a boiler, in the kitchen to supply hot water. Generally only one fireplace was ever lit, in the main living room, and the kitchen boiler ran continuously.

It was an onerous task to clean out the living room grate and relight a fire every day for seven months of the year, but it was soon to come to an end in the 1950's. What people did not realise was that the chimneys on all of their houses exhausted to atmosphere less than 20 m above the ground.

London was noted for its fogs to the extent that they featured in literature and popular songs. But in November 1952 a temperature inversion over the southeast of England created a thick blanket of smog over London and its suburbs which led to the deaths of more than 4000 people from bronchial ailments. This so alarmed the Government that in 1954 they passed the Clean Air Act which effectively brought an end to the burning of bituminous coal in domestic grates.

Coke and anthracite, which is almost pure carbon, could still be burned, but the passage of the act brought about the end of the domestic market for coal. It took several years for this to happen, and it led to the wider introduction of central heating with oil- or gas-fired boilers. With coal, gas and electricity all in public ownership there was no incentive to introduce district heating. The first gas was discovered in the southern North Sea in 1964.

Natural gas replaced coal gas in the domestic market, and diesel-electric and electric locomotives replaced steam on the railways. Coal mining ended in Belgium and the Netherlands. Many new steam plants of the time were designed for oil firing and some of the older plants were converted from coal to oil, and it was three of these, in Belgium, Ireland, and Croatia which were later repowered with gas turbines to create some of the first combined cycles in Europe.

But coal has fallen out of favour in much of the developed world and European production is much less now than it was. In the present European Union of twenty-seven countries only Poland, Germany, the United Kingdom, and Spain have significant coal production, and there are few plans for new coal-fired generating capacity being built across the region. The big markets for coal in power generation are now China, India and the United States.

Germany has sub-bituminous coal in the Ruhr Valley and large lignite deposits in the Rhine Valley above Cologne. Thirty years ago the Government put an extra charge on the cost of electricity (the so-called coal-pfennig) to support the industry. In the east there is another large deposit of lignite extending from Cottbus in the north down to Leipzig, which was the principal fuel for the power stations of the former German Democratic Republic.

The coal industry of the United Kingdom has shrunk from 850 pits at nationalization in 1947 to fewer than twenty today. Some of it was due to geological problems and some to loss of market due to trade union militancy in the seventies and early eighties. More than 75% of the resource is still in the ground and because of the high sulphur content of some of the most accessible deposits, owners of the oldest power plants,

4.1: Hong Kong: Castle Peak B station with 4 x 670 MW coal-fired sets is being fitted with SCR and FGD sytems to enable it to continue running after 2011.

which must shut down by 2015 at the latest, preferred to import low-sulphur coal rather than fit Flue Gas Desulphurization (FGD) systems for the remaining few years of operation.

As a fuel for power stations it has been burned inefficiently. Most of the coal-fired power stations around the world operate on a low-pressure subcritical steam cycle, with at best, a thermal efficiency of 35%. As environmental issues, particularly acid rain, caught public attention, new coal-fired power plants had to have low-emission burners, and FGD, in addition to electrostatic precipitators which collect dust carried over in the flue gases.

Today, older units such as, for example, Castle Peak B in Hong Kong, must now have these environmental measures installed if they are to continue in operation. This station was completed in 1990 with four 670 MW coal-fired sets and is now the subject of a major environmental programme to remove nitrogen and sulphur oxides from the flue gases so as to enable it to stay in operation after 2011.

Similar measures have been or are being installed at other coal-fired plants around the world to ensure that they can keep running. However none of these measures enhance the performance of the power plant, and all add to its cost.

Although all these measures are now mandatory for new coal-fired power plants, they have had an effect on performance by accelerating the introduction of the supercritical steam cycle with its higher pressure and temperature offering a higher efficiency. Another ten percentage points made it worthwhile to build a coal-fired power plant with all the environmental measures in place.

So a new coal-fired power plant ordered today would be a more compact system and of higher efficiency than those built more than forty years ago. Then a typical power plant would have up to four generating sets of between 300 and 660 MW and with a multi-flue stack up to 200 metres tall to disperse exhaust gases on the prevailing wind. The only environmental measure would be the provision of either electrostatic precipitators or a bag filter house to collect dust carried over in the flue gases.

In the production of supercritical steam there is no discernible phase change at the boiling point. The density of the steam and the evaporating water are the same. The condition has been known for over 100 years. Mark Benson's original 1922 patent for the once-through boiler said that a long tube containing supercritical steam would produce dry superheated steam if a valve on the end of the tube were to be opened.

The problem in Benson's time was the failure of high pressure drums, which were then made of riveted plates, and had been known to explode, often with disastrous results. His idea was that there should be no drum but a continuous high-pressure tube fed from a circulation pump at one end and with the steam valve at the other. This is the principle of the once-through, or Benson boiler.

Once-through, boilers have been applied to subcritical steam cycles, particularly for plants designed for mid-load and peaking duties which would have rapid starting capability through not having to cope with the thermal inertia of a large high-pressure drum. But with the introduction of supercritical steam cycles the Benson boiler has come into its own.

Supercritical steam had to wait for metallurgy to catch up. Some early plants used austenitic steels for the highest temperature components, which were expensive to produce. But the development around 1980 of stronger ferritic steels which could support supercritical pressures at lower temperatures made it possible to introduce steam pressures at over 200 bars and achieve a gain in efficiency to more than 40% at steam temperatures which were not much greater than those of the existing subcritical steam cycles.

Supercritical steam has given coal a new lease of life in some of the producing countries, but the environmental measures, principally FGD and low NOx burners with Separate Over-Fire Air (SOFA) to further reduce emissions also reduce some of the efficiency gain of the higher-pressure steam cycle.

TABLE 4.1: SOME COALS FOR POWER GENERATION

Source	Type	LHV kJ/kg	Water %	Ash %	C %	S %
Australia						
Bayswater	Bituminous	25143	10.0	12.5	64.4	0.8
Hunter Valley	Bituminous	27905	4.8	9.4	70.7	0.6
Colombia						
El Cerrejon 1	Bituminous	25947	9.0	9.6	66.5	0.7
El Cerrejon 2	Bituminous	26705	8.6	8.8	67.9	0.8
BP Columbia	Bituminous	28267	6.6	9.1	70.9	0.8
Germany						
Rheinbraun	Sub Bituminous	22100	11.0	4.0	58.8	0.4
Hambach	Sub Bituminous	19273	18.0	3.5	52.4	1.1
Lausitzer	Lignite	8801	55.8	3.9	27.0	0.8
Poland						
Weglokoks	Bituminous	25117	9.2	13.6	65.5	0.5
Janovice	Bituminous	29152	3.5	5.6	74.7	0.4
Russia						
Kuzbass 1	Bituminous	30645	5.0	5.1	78.6	0.3
Kuzbass 2	Anthracitic	31088	4.5	5.0	81.9	0.3
South Africa						
Greenside	Bituminous	25821	8.0	12.7	65.5	0.5
Kleinkopje	Bituminous	24733	8.1	14.7	65.5	0.6
Douglas Steam	Bituminous	25540	7.5	12.9	67.1	0.5
Koornfontein	Bituminous	25435	8.0	13.4	66.2	0.4
United Kingdom						
Bilston Glen	Bituminous	26574	11.9	5.8	68.5	0.8
Selby	Bituminous	23331	10.8	17.3	59.4	1.2
Thoresby	Bituminous	24625	10.2	13.4	62.9	1.6
Drakelow	Bituminous	23961	12.2	11.4	61.0	2.4
United States						
West KY 11	Bituminous	17837	10.3	31.8	44.6	3.4
Pittsburgh 8	Bituminous	27680	6.0	9.9	69.4	2.9
Illinois 6	Bituminous	22325	12.0	16.0	55.4	4.0
Rosebud, MT	Sub Bituminous	19108	25.2	8.1	51.5	0.6
Texas Lignite	Lignite	14880	37.7	6.5	41.3	0.7

Performance depends also on the quality of the coal which varies from one field to another and from one country to another. There are wide variations in calorific value, carbon, ash, moisture and sulphur content. In the past coal-fired power stations have been built close to coal fields if not at the mine mouth, and the design of the plant is basically tailored to the properties of the available coal supply.

4.2; Schwarze Pumpe, Germany, 1600 MW supercritical steam plant, also sends steam loads to a briquette factory and the local district heating network. (Photo courtesy of Siemens)

Several countries with notably low-sulphur coals have developed trade with Europe for the old power stations which have not fitted FGD and are on limited running hours under the LCPD until they close at the end of 2015. Russia's Kuzbass region, for example, has coals characterised by high calorific value and extremely low contents of ash, moisture and sulphur.

A typical example of a modern supercritical steam plant is the power station at Schwarze Pumpe, 30 km south of Cottbus. It has two 800 MW turbogenerator sets which entered service in November 1997 and March 1998. It was designed for base-load operation with steam take-off to a briquette factory next door, and to a district heating network serving the villages of Schwarze Pumpe and Spremberg.

Steam conditions are 260 bars 560°C with reheat at 53 bars 565°C. This compares with 160 bars, 540°C for a subcritical steam cycle. The turbine has a single-flow high-pressure cylinder, double-flow intermediate and two double-flow low pressure cylinders with an underslung condenser.

The FGD system comprises two streams per unit after the electrostatic precipitators which remove dust from the flue gases. The gas then rises up an absorption tower through a quench spray and meets an alkaline spray coming down. Sulphur dioxide in the flue gas reacts with calcium carbonate suspended in the spray, after which the cleaned gas goes through a moisture separator to the stack. The FGD units produce in

4.3: Zheijang, China: One of four 1000 MW supercritical steam turbines in the Yuhuan power station of Huaneng Power Inc. (Photo courtesy of Siemens)

total some 500 000 t/year of gypsum which is sold to a local factory making plaster board.

Supercritical steam is the current technology, with over 20 years of operational service in Europe and the Far East. The efficiency at Schwarze Pumpe, burning the locally mined Lausitzer lignite is 41% in pure condensing mode and 55% in full combined heat and power mode, with 800 t/h of steam at 4.5 bars going to the briquette factory, and 80 MJ/s at 110°C fed to the district heating network. A power station such as this, if ordered in 2010 could be in operation by the end of 2015.

With the introduction of supercritical steam a number of improvements have made coal-fired power plants more environmentally friendly. First, higher efficiency from the improved steam conditions gives a reduction in exhaust gas emissions of at least 20%, and most probably more, given the likely age and efficiency of the plants that they would replace. FGD produces a saleable commodity in gypsum which can be sold to make plaster board. Higher combustion temperatures and low NOx burners with SOFA result in an ash containing less unburnt carbon for which there are more opportunities for sale as a construction aggregate.

The immediate future of coal is with supercritical steam plant but there are other concepts which are further into the future and have yet to be proven, first that they work effectively and second that they offer an improvement in performance over the current systems which, at the present state of the art, they cannot.

So supercritical steam, in its latest version, is the most efficient coal-fired plant design so far. The higher efficiency the less coal has to be dug and transported to site and there is less ash to be removed.

The first ultra-supercritical steam plant with high pressure output at 262.5 bars, 600°C has gone into service in China with Huaneng Power Company. There are four 1000 MW units, two of which started operation in 2007. with the others following a year later. This is the ultimate supercritical technology at present and several plants of this type are planned in China and elsewhere.

This power plant was supplied by Siemens, but in April 2008 Alstom picked up an order in Germany from RWE, to supply boilers for two 800 MW units going into a new power plant at Hamm. These will be their first with the ultra-supercritcal steam cycle at 260 bars, 600°C and the company is forecasting an efficiency of 46% which compared with 36% on the best subcritical cycles means that it will burn 23% less coal than the subcritical plant of the same output and with correspondingly lower emissions.

These plants will make a significant improvement to cleaning up coal technology. For that reason these could be the definitive coal-fired power plants for the foreseeable future if they have to be built at all. But for how long do we have to wait for a proven clean coal technology which does not destroy the performance gains with a high auxiliary load and an ultra high price for the electricity to the consumer, because if this is all that we can expect then it will be publicly unacceptable.

Several coal producers have investigated the concept of underground gasification which would entail drilling into a coal seam to receive an oxidizing gas mixture which would cause it to react and produce a combustible gas, mainly hydrogen and carbon monoxide, which would be sent up to a power plant on the surface.

All the ash remains underground but any sulphur would be separated out on the surface. More important, there would not be the additional lifetime energy loads involved in mining the coal and distributing it to the power stations. The process has been demonstrated on a small scale but development costs would be high, particularly with deposits under the sea and a long way from the mine shaft.

Coal gasification is a long established process in the coal producing countries of Europe and North America. Coal gas dates from the nineteenth century as a public energy supply in the major cities, initially for street lighting and later extending to cooking and space heating in the domestic market. The processes that created it were the building blocks of the chemical industry and of the coal gasifiers that have emerged a hundred years later. Older people will remember the large cylindrical gasholders at their local gasworks which by now will have been long demolished with the introduction of natural gas.

The integration of a coal gasifier with a combined cycle was first proposed in the early 1970's. In Germany the Kellerman Lünen plant of the utility STEAG, completed in 1972 was one of a series of fully-fired combined cycles built around that time in which the gas turbine exhaust was fed to the burners of the boiler supplying a large steam turbine. On its way the exhaust gas passed through a series of heat exchangers which replaced the turbine's standard feed water heaters. Where the Kellerman Lünen plant was different was that instead of natural gas, the gas turbine burned synthetic gas produced by a Lurgi gasifier on a neighbouring site, using locally-mined bituminous coal..

At the time, this was one of several schemes to improve the efficiency of, in this case, an oil-fired power station. Efficiency was 45% using a 50 MW gas turbine with a 350 MW oil-fired steam turbine. Only two plants of the eighteen which were built had coal-fired steam turbines. The arrival of larger gas turbines running at synchronous speeds steered application towards the purpose built, gas-fired combined cycle and later the integrated gasifier and combined cycle (IGCC) which is now approaching commercial application in the United States.

IGCC is a power plant, linked to a chemical processes. The basic elements are an air separation plant, which produces oxygen for the gasifier, the pressurized reactor vessel of the gasifier, and synthetic gas clean up, including sulphur and mercury recovery systems. Besides electricity there are saleable products in the recovered sulphur and mercury, and of course the gasifier ash.

IGCC has the largest auxiliary load of any of the generating systems, which reduces the net saleable output. The basic efficiency measured on the generator terminals depends on the extent to which the combined cycle can recover energy from the syngas coolers. But the net energy efficiency, defined as the ratio of saleable energy to the fuel energy input, gives a worse result than current best results obtainable with coal-fired supercritical steam plant, and this before any consideration is given to the economics and performance of carbon capture.

Of the various gasifier types available the most widely used is the General Electric (GE) Texaco system, a slurry-fed oxygen-blown gasifier used in nine of the 23 schemes that were in operation around the world at the end of 2008. This gasifier was used in the first IGCC scheme at Cool Water, about 150 km northeast of Los Angeles, which was completed in 1984. But this was a prototype plant with the gas

4.4: Wabash River, IN, USA. 250 MW IGCC went into operation in 1997 snd in separate campaigns has used high sulphur coals and petroleum coke feedstocks. (Photo courtesy of PSI Energy)

turbine of the day. It was the first example of a coal gasifier which was linked to a combined cycle consisting of a GE Frame 7EA with a 40 MW steam turbine on a separate shaft. At the time the cost of the complete plant was put at $2500/kW installed. Since then costs have come down and commercial plants planned for service after 2012 are expected to be at about $1800/kW installed.

Commercial IGCC schemes have for some years been operating in the oil industry using residual oil and petroleum coke as feedstocks. The combined cycle is built as a combined heat and power scheme supplying steam to the refinery processes. Of the ten schemes all except two use the Texaco oxygen-blown gasifier. The plants at Shell's Pernis refinery in the Netherlands, and at AT Sulcis in Italy, have the Shell gasifer which is an oxygen-blown dry-fed system.

Five of the schemes are in Italy, two in the United States and one each in the Netherlands, Singapore and Japan. All are combined heat and power schemes supplying power and process steam to the refineries. But why there are only eight such schemes? Twenty years ago, Shell saw gasification in the oil industry as a potentially large market which has never come to fruition.

Italy, with five schemes is understandable because it has to import all its oil and gas, has no coal and abandoned nuclear power for blatantly political reasons to solve a government crisis in 1988.

TABLE 4.2: IGCC SCHEMES AT END OF 2008

Site	Country	Gasifier	Feedstock	MW
Kellerman Lünen	Germany	Lurgi	Coal	163
Cool Water	USA	Texaco	Coal	110
Buggenum	Netherlands	Shell	Coal[1]	253
Wabash River	USA	Destec	Coal[2]	250
El Dorado	USA	Texaco	Pet. coke	42
Polk County	USA	AFB	Coal	107
Puertollano	Spain	Prenflo	Coal[2]	300
Pinon Pine	USA	AFB	Coal	110
Shell Pernis	Netherlands	Shell	Resid. oil	110
Falconara	Italy	Texaco	Resid. oil	220
Priolo	Italy	Texaco	Asphalt	500
Sarlux	Italy	Texaco	Resid. oil	500
Schwarze Pump	Germany	Lurgi/GSP	Lignite[5]	75
Delaware City	USA	Texaco	Pet. coke	235
Jurong	Singapore	Texaco	Cracked tar	160
Negishi	Japan	Texaco	Asphalt	400
Vresova	Czech Rep.	HTW	Coal	358
A T Sulcis	Italy	Shell	Pet. coke	456
Sanazzaro	Italy	Texaco	Resid. oil	250
Plaquemine	USA	Destec	Coal	208
Sanghi	India	GTI UGas	Coal	110
Clark County	USA	BG Lurgi	Coal[3]	540
Iwaki	Japan	Mitsubishi	Coal[4]	250

[1] Mixture of coal and biomass. [2] Mixture of coal and petroleum coke. [3] Coal and pelletized refuse. [4] Air-blown gasifier. [5] Two gasifiers for plant incorporating combined cycle.

Japan, likewise, imports 80% of its fuel supplies and seeks to improve energy efficiency. Although it used the Texaco oxygen -blown gasifier for the Negishi refinery scheme, the builder, Mitsubishi, at Iwaki, are testing an air-blown gasifier for power generation applications.

The coal IGCC projects to date have all been demonstration schemes supported by EU and US Government funding. They were viewed as prototype systems for eventual application to electric power generation. IGCC was viewed as clean coal technology since it was a coal derived gas cleaned of the trace elements before combustion; and with an inert granular ash which can be sold as a constuction aggregate.

At the present time it can be said that IGCC is proven technology but not at a price and performance that would be attractive to an electric power generator. As a power plant there is still a lot of development work with the gasifier and the combined cycle before it will be competitive, and if carbon capture is included this could add too much to the cost of construction and the price per kWh of the electricity produced.

Coal gasifiers have been used in the chemical industry to produce feedstocks and are still in demand. But to link the gasifier to a combined cycle power plant there are a number of issues still to be resolved, not least in the performance of the integrated system. As an independent gasifier supplying synthetic gas to the chemical industry the reliability and maintainability of the gasifier is well understood. But a much larger scheme linked to a combined cycle and operating for over a year at a time, and at varying levels of output, is a totally different issue.

IGCC has four connections between gasifier and power plant: electric power to drive the gasifier, synthetic fuel for the power plant, nitrogen from air separation for flame dilution and to improve mass flow, and steam from the gasifier and syngas cooling systems, to the heat recovery boiler.

The synthetic gas is produced at about 750°C as it leaves the gasifier. and has to be cooled to 50°C for the gas clean-up process which removes sulphur and and mercury. So there are two sources which can be used: the jacket cooler of the gasifier vessel and the syngas coolers, which are a weak spot of reliability with a dirty product gas on one side and clean, high-pressure feedwater on the other. But it is the extent to which these heat sources can be integrated to the combined cycle that determines its efficiency.

Development of IGCC is coalescing around the gas turbine majors. Three of them now have their own gasifier technologies. GE acquired the Texaco gasifier technology in 2004. In Japan, Mitsubishi has developed an air-blown gasifier of which a 250 MW prototype plant has been built at Iwaki and is now on test.

In 2006 Siemens took over the GSP gasifier technology from the Swiss Sustec Group. Their first project will be an independent gasifier at Spreetal, Germany, to produce synthetic gas for conversion to sulphur-free diesel fuel. The interest in IGCC is currently with the Conoco Philips E-gasifier for which they have designed a 630 MW class IGCC plant based on the SGT6-5000F gas turbines for the North American market and usable with lower rank coals and lignite.

After GE acquired the gasifier technology from Texaco they entered an agreement with Bechtel to produce a standard IGCC system with a combined cycle using the Frame 7FB gas turbine.

In 2005 GE and Bechtel received orders from American Electric Power (AEP) to perform feasibility studies for a planned 630 MW, IGCC plant on a site in Meigs County, Ohio, and a second identical plant for a site in West Virginia. Later Duke Energy also placed a design contract for a 600 MW plant to be installed at Evansport, Indiana. These initial design contracts were completed at the end of 2006.

IGCC is future technology but there has up to now been little interest among electric power generators and certainly not outside the United States. But since these first projects were announced in 2005 some 50 schemes have been proposed in the United States and another 26 in the rest of the world, but as more information has come to light many of these have not been followed up.

Undoubtedly the global warming scare has revived interest because it is now being promoted as clean coal technology. But only if it can be developed to the point that steam production from the cooling of the syngas can be optimised, that efficiency is at least as good as that of the best supercritical steam plants, and reliability is as good as that of the best combined cycles.

That is certainly not the case at present, without the addition of carbon capture and storage. But the future of coal is now tied up with the global warming issue. If global warming is due to carbon dioxide being poured into the atmosphere then the answer is to take carbon dioxide out of the flue gases. If it has to be done at all then it is easier to remove it from the synthetic gas before combustion than from the flue gases after combustion.

This would seem to be a somewhat extreme measure and can only have been thought up by somebody who doesn't believe in emission-free nuclear power, and claims that coal is the fuel of the future to generate our electricity. In other words it is an idea of the Green Movement.

Yet is it feasible, and what would have to be done to bring about a significant improvement in the environment? Development of clean coal technology has received funding from the US government as announced in the 2006 State of the Union message. The aim was to assist the development of the technology and in particular of carbon dioxide capture.

For IGCC, the carbon dioxide capture process is a high temperature shift reaction of the carbon monoxide in the syngas with steam which produces hydrogen and carbon dioxide. It is proven technology in that it is used in the ammonia industry. But that is on a much smaller scale and if integrated as part of a 700 MW IGCC scheme it would be a continuous process to separate the carbon dioxide which could be sent to a nearby oil field where it can be pumped into the well to enhance recovery of the oil.

There may be industrial uses for the carbon dioxide, but for this, the gas would be recovered from an industrial boiler and in quantities of less

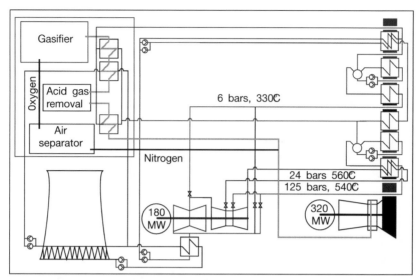

4.5: Typical arrangement of IGCC scheme using F-class gas turbine with optimised cooling arrangement, but without a carbon capture system for the syngas.

than 1 million t/year. For a coal-fired power station, the amount of gas to be captured is about 1 million t/year per 100 MW of output capacity.

But at the current state of the art, without carbon capture, the planned schemes for the United States market still have a long way to go to catch up with the performance of supercritical steam. The GE/Bechtel design with two Frame 7FB gas turbines and a steam turbine, and without carbon capture, has an output of 640 MW at an efficiency of 39%, compared with the normal 207FB combined cycle on natural gas at 562.5 MW and 57.6% efficiency.

The Siemens design based on the E-gasifier is rated 612 MW at 38% efficiency compared with the standard combined cycle at 589 MW and 57.5% efficiency. The two systems differ in that the GE design is only for high-grade bituminous coals, whereas the Siemens design can handle sub-bituminous coals and lignite which might give it a wider market.

The net efficiency is calculated on the same basis as for any combined cycle: from fuel to saleable electricity. The heavy auxiliary loads of the gasifier and the gas cleaning system are responsible for the low net efficiency. On the other hand it is a clean fuel which enters the gas turbine with no sulphur and no mercury, as does natural gas anyway.

To improve the efficiency the first change is to improve cooling of the syngas which would increase the steam flow to the turbine. Second would be to use the H-class gas turbine which has an inherently higher efficiency and output, with the GE Frame 7H at 260 MW and the 50 Hz Frame 9H at 330 MW on natural gas and the similarly rated Siemens SGT 5-8000H.

Whatever the steam pressure, the syngas cooling would have to be with a two-stage heat exchanger from syngas to subcritical steam, this would reduce the likelihood of corrosion in the coolers causing the syngas to leak into the steam path, because it would be detected in the intermediate section and force a shut down before it could do any damage to the steam circuit. The high steam pressure would be a further protection.

If as a result of these measures we can see an IGCC efficiency nearer 50% then there might be a clear advantage over coal-fired supercritical steam, which will be worth having for the 14 % efficiency advantage over old coal and the lower carbon dioxide output that it would offer. But for the time being development has stalled.

The emphasis is moving to carbon capture both for IGCC and existing coal-fired steam plants. Like all the other environmental additions, it reduces the output of the plant and increases the cost. Furthermore the combined cycle output and efficiency will be lower on hydrogen fuel and the gas turbine burners have to be modified to work with it and also be multi-fuel devices to allow for starting and shutting down on natural gas or distillate.

No carbon capture system has been coupled to a gasifier integrated with a combined cycle to see how the complete system would operate, and in any case it may not be until 2015 before there is any worthwhile experience with IGCC in commercial operation, without carbon capture, to see if it is worth doing.

But is this experience likely to happen? The feasibility studies were completed at the end of 2006 but nothing has been carried forward to actual project engineering. In fact IGCC performance has to be improved before it can be considered a worthwhile investment for any power generator. Furthermore studies of carbon capture systems based on existing practice in the oil and gas industries suggest that they will make it even less attractive.

Without carbon capture the auxiliary load is about 130 MW for an IGCC with a gross output of 770 MW. Apply carbon capture and it drops to 745 MW because the gas turbines have a lower output when burning only hydrogen. The additional load of the carbon capture equipment is 60 MW and the net output of the combined cycle is therefore 555 MW. More important is the fact that this saleable output is 27.9% less.

If this is the best that we can expect from IGCC with carbon capture it negates everything that has been done to improve the efficiency and cleanliness of emissions of coal-fired power plant for the last fifty years. What is more to the point, who is insisting that we take all the carbon dioxide out of the flue gases of every thermal power plant when, for base load power at least, there is a proven system of generation with forty years operating experience and no emissions of any sort.

The other issue with carbon capture is what do we do with the recovered gas, and what effect will that have on the siting of the power plant. If the gas can be sent to an oil field to be used for enhanced recovery more oil would be produced and the gas would be locked in the oil-field, and it would therefore have a market value. Therefore the power plant should be built near the oil field. If the gas is to be treated as a waste product as undoubtedly most of it will be, then where do we put it, and who pays for it to be done?

Saline aquifers which are found in many parts of the world are one solution. But is there sufficient capacity of these aquifers for all the carbon dioxide storage that would be needed. Could an aquifer take all the carbon dioxide produced from a large coal-fired power plant, say 20 million t/year for the whole of its 40 year life?

This is the problem with coal which is going to get worse with time. If certain governments are unwilling to grasp the nuclear nettle then coal it must be. The Greens have in the past taken delight in trying to prove that the energy used to build a nuclear power plant is more than it would generate during its operational life. They have not done a similar exercise with a coal-fired plant but it could be a lot nearer to the truth given the distance from mine to power plant and the energy cost of the mining operation, the transport of the coal to the plant and the removal and disposal of the ash.

The 1000 MW ultra-supercritical plant at 46% efficiency would burn about 50% less coal than two IGCC with carbon capture at 32.5% efficiency for the same electrical output. In other words two IGCC with carbon capture would produce the same power as one supercritical steam plant without it.

Then the electric power supplied to the railway bringing the coal from the mine wherever it is would also be about 25% less to the steam plants than to the two IGCC. So for every four trains delivering coal to the steam plant there would be five trains delivering to the IGCC's, which is another energy load attributable to carbon capture.

Then, assuming that carbon capture works the pipelines have to be constructed to the oil field or the repositories, together with boreholes to place the gas and the equipment to compress the gas and control the flow. Also if carbon capture becomes a firm condition of operation, the gas pipeline must be available when the plant enters service. If there was any breakdown on the compressors, so that the gas flow was reduced or stopped altogether, would the power plant then be allowed to operate?

The existing power stations, which will not have carbon capture, supply all the energy to produce the steel and manufacture the pipelines. Add this to the energy used in building five IGCC which are a group of chemical plants, one for each gas turbine in the combined cycle that they are serving.

But there are more disadvantages: An IGCC plant of about 550 MW with carbon capture and an efficiency at about 32.5%, is at a similar level of performance to that of the coal-fired plants built in the 1950's. These had subcritical reheat steam cycles and bag houses or electrostatic precipitators to collect dust caried on the flue gases, which were the only environmental measures at the time.

There were worries then about the efficiency of power generation, which, in the United States were solved by adding one or other of the gas turbines of the day with a heat recovery boiler replacing the feedwater heaters which were normally fed by bleeds from the steam turbine. With the bleeds shut off, the turbine could generate more electricity to which would be added the output of the gas turbine.

Let us look into the future to a time when carbon capture has been integrated to an IGCC and the performance and costs are known, which may not be until around 2020. By this time, too, the H-class gas turbines will be proven in service. So a generating company has the choice: a 530 MW combined cycle burning natural gas at an efficiency of 60% ; or the 530 MW hydrogen-fired combined cycle on the back of a gasifier at 32.5% efficiency but with 90% less carbon dioxide emission and 50% higher generating cost.

The important issue here is the saleable output of electricity because this is how the generator will earn the money to pay for the plant. In a competitive merchant plant environment, the generating company will undoubtedly go for the combined cycle with almost double the efficiency, and less than half the emission levels. Cost is one factor; the cost of construction, which is determinate at any given time; and cost of fuel which in recent time has varied widely up and down.

All coal-fired power plants are considered candidates for future carbon capture, whether in the fuel treatment, as with IGCC, or post-combustion from the stack gases as for a coal-fired steam plant. But no full scale system has yet been demonstrated with a power plant, and the deterrent is surely the cost of the plant and the lower output.

So the coal-fired power plants of the future are with each additional environmental measure becoming less efficient and more like chemical factories with an electricity byproduct. Is this really what the electricity generators want to be? There would be no such considerations with the gas-fired combined cycle, although the price of gas, as has happened already, would determine whether the plant would run base-load or only at peak times, or be shut down and mothballed until the price falls.

So the choice of what to build is clearly in favour of the combined cycle, not only because of its high efficiency, but also its low environmental impact and the fact that it could be up and running in two years from the date of receiving consent to build.

The situation today then is that we can generate a gas from coal and burn it in a gas turbine. But we cannot strip out carbon dioxide to convert it to hydrogen. We know how to do it but nobody yet has attempted to apply the system to an IGCC scheme in continuous operation.

Is coal then a firm option for the future or should it be taken out of power generation altogether? With so much electricity generated by coal in the world, that may not be easy, but there are in Europe and the United States large inventories of old coal-fired capacity, most of which is still running. It would mean that as existing coal-fired power plants are shut down and demolished, they should be replaced with gas-fired and nuclear capacity. But the countries with the longest history of coal firing in Europe and North America initially would be able to close the greatest number of plants. Where new plants have been built to replace old coal-fired units, these have been mainly combined cycles.

Can there really be a case for stripping out the carbon dioxide from fuel, or flue gases when there would be a huge increase in the supporting energy used to set up the system? If a pipe has to be laid from a plant on the North Sea coast of the UK to carry carbon dioxide out to an oil field, then we are looking at a steel pipe perhaps perhaps 2 metres in diameter by 400 km long, which has to be manufactured and installed, along with the compressors needed to drive the gas flow to the platform.

One can determine the amount of carbon dioxide that would be produced and injected during the 40 year life of the power plant. But what would be the gross national output of carbon dioxide from all sources associated with it? It requires energy to mine the coal, and transport it to the powerplant and also limestone for the FGD system. Then there are the ash and sulphur products which must be taken to the end users.

These larger lifetime supporting energy loads are inseparable from coal energy in a way that is not the case with all other energy systems.

It may be clean coal with low emissions but as an exercise in energy conservation it is a disaster. Only the removal of coal from power generation will see a drop in energy consumption, and of greenhouse gas emissions overall.

One measure which will bring about the closure of a lot of coal-fired capacity in Europe in 2015 is the European Union Large Combustion Plant Directive (LCPD). This is aimed to reduce the acidification of rain, ground level ozone and particulates by controlling emissions of sulphur and nitrogen oxides and dust from large combustion plant. It applies to all electric utilities within the 27 countries and classifies their coal-and oil-fired power plants as to when they were built.

All plants built after 1987 must comply with the Directive to stay in operation after 2015. Many of them will have been supplied from new with FGD systems and the other environmental measures. Plants built before 1987, if they were not brought up to standard by the end of 2007, which meant basically fitting FGD, would have to opt out from the Directive but can continue in limited operation, about 2500 h/year, until 2015 when they must shut down. On this basis, and given that the only other coal-fired plants remaining in Europe would be those built between 1987 and 2015, some plants could still be running in 2050.

For an existing coal-fired steam plant, post-combustion carbon capture is the only practical solution to cutting emissions, but it is still in the early days of development. Since February 2008 a system has been on test in the United States at a coal-fired power plant in Wisconsin. This is taking a slip stream off one of the flues equivalent to 20 MW of generation. In Europe similar tests are planned for some of the stations which have opted out of the LCPD. These stations could serve as research facilities until they shut down to determine how successful is post-combustion carbon capture, and whether it can be applied to a new supercritical steam plant.

It is only 20 years ago that the United Kingdom got 80% of its electricity supply from coal. The Government Energy policy announced in January 2008 put emphasis on new nuclear build because of the need to replace so much generating capacity after 2015, while achieving at the same time an 80% reduction in greenhouse gas emissions by 2050.

Of the coal-fired plants, the last to be built was Drax, completed in 1985, and which provides approximately 7% of the electricity supply of the UK. Owned and operated by Drax Power Ltd the station was built in two stages with three sets completed in 1976 and a further three added in 1985 All were subsequently fitted with FGD systems which on full load can remove 280 000 t/year of sulphur dioxide which is made into

4.6: Vresova, Czech Republic: gasifier makes clean transport fuel with no sulphur. Is this the real future for coal gasification? (Photo courtesy of Siemens)

gypsum in the process, and which is sold to British Gypsum to make plaster board.

Drax Power are currently undertaking two programmes which will reduce their carbon dioxide output by some 3 million t/year. All six turbines are getting upgrades on all stages which will add the equivalent of another set and raise the efficiency to 40%. That will reduce carbon dioxide by about 1 million t/year and the other 2 million t/year will be taken up by co-firing with solid biomass.

In May 2008 the company awarded a contract to Alstom for the biomass processing plant. This will receive 1.5 million t/year of various biomass and processs it for direct injection into the power station boilers. Typical biomass would be sawdust and wood chips from sawmills, coppiced willow, and commercially grown miscanthus. Biomass can provide 10% of the fuel and be able to generate about 66 MW of the total output of each set for a total of 400 MW on which they would get the renewable energy payment that can be set against the cost of their coal.

Drax is not the only plant which is co-firing biomass. It was the first British plant to have FGD installed and is opted into the LCPD. It was the last coal-fired power station to be built in the UK and may be the last to close, hence the emphasis on reducing carbon dioxide emissions. They can do nothing else since there is no proven system of carbon capture and sequestration available.

4.7: Drax, UK. This 4000 MW coal-fired power station, will from 2009 be co-firing 10% solid biomass along with turbine upgrades to reduce greenhouse gas emissions.

Several of the 2000 MW stations are also co-firing up to 5% biomass, particularly those stations which have opted out of the LCPD and which must close at the end of 2015. Of the twelve 2000 MW plants built since 1965 five have added FGD but since these are now all over 40 years old this was probably to keep them running until new plants could be built. One, an oil-fired plant at Pembroke, South Wales, is among a number that have already been demolished. EdF Energy expect to have shut down their two plants at Cottam and West Burton by 2020.

Another 3200 MW already have been demolished at High Marham, West Thurrock, and Drakelow. Combined cycle are to be built at Pembroke (5 x 400 MW) and Drakelow (4 x 400 MW). A third former coal-fired plant which was planned for a combined cycle more than ten years ago is at Staythorpe, a few km north of Newark on Trent. The new plant has four 400 MW single-shaft combined cycle blocks.

Kingsnorth on the Isle of Grain is earmarked for a 1600 MW coal-fired ultra-supercritical steam plant to replace the existing 2000 MW coal-fired station there which must close in 2015. The plan has been fiercely criticised by the Green lobby which brands it as dirty coal energy replacing dirty coal energy.

In April 2009, the British Government announced plans for five large coal-fired power stations which would only be built if the developers could include a proven CCS system, or allow for its future installation.

The final decision as to what is built rests with Governments, they have accepted the principle of the LCPD but all European Union Governments must face a General Election before 2015, and it is unlikely that the present parties will all be in power by then. Energy is not generally an issue to win or lose elections. But as shown in Germany ten years ago, the extent of Green support in government can have a profound effect on future energy policies.

One country has already spelt out an end to coal on grounds of public health. In Canada, in the Province of Ontario, the Provincial Government's energy plan of July 2005, announced the closure of the six coal-fired plants in the province and their replacement with two nuclear stations and a number of combined cycles for mid load and peaking duty. Already the two plants, Hearne and Lakeview, in Toronto have been demolished and a 550 MW combined cycle has been built on the Hearne site.

This is the first example of the complete removal of coal from a power system though there is some doubt as to when the closures will be complete since the nuclear sites have not been finalized and the combined cycles have to be built and some of the older hydro plants upgraded. The six plants were widely spread out with the two largest, Lambton, near Sarnia, and Nanticoke on Lak Erie, built in the late 1960's at a time when the province was diversifying from an all hydro system to a mixed hydro, nuclear and coal system, even if it meant bringing coal in by rail from Nova Scotia or importing it across Lake Erie.

Elsewhere in Canada, mainly in Alberta and Saskatchewan, coal-fired plants, as in the UK, will only be approved if they can include provision for CCS.

Given what has since come to light from studies carried out by the US Department of Energy, and others it may not be possible to reduce the cost of the process and its integration with the power plant, but then there is the increased auxiliary load not just of the power plant itself, but all the additional energy loads of bringing more coal to a less efficient power plant and sending the separated gas to a distant oil field.

So the question that we have to ask is why is so much effort being made to make coal-fired power clean and emission free with a system which will reduce the efficiency and add substantially to the price of electricity, and more important, add significantly to the energy loads of the supporting services, and construction?

A new power plant with carbon capture and all the other environmental measures may be hopelessly uneconomic. To build one now with provision for later fitting of carbon capture is at best dubious because we don't know what system could be used and how efficient it would be and how much steam would have to be taken from the turbine.

Look at another case. A 2000 MW coal-fired plant completed in 1970 is fitted with FGD so that it can keep running after 2015 when it will have been running for 45 years at an efficiency of 35%. Gypsum sales will have helped bring a quick payback for the FGD system, but maintenance costs will be steadily rising and the completion of a 1600 MW nuclear plant in 2017 could be the signal for it to close down.

So coal is gradually losing the battle. There are research projects to develop and apply carbon capture, but unless the basic performance, in terms of reliability and efficiency, can be improved at a much lower cost than at present, there is no real future for it. Secondly much of the coal-fired capacity in the world is old and of low efficiency and will have to be replaced long before a viable large-scale carbon capture scheme has been developed.

The answer then must to be to take coal out of power generation globally, starting with Europe and North America where there is the greatest concentration of old and inefficient coal-fired generating capacity. If we go back forty years to a time when there was genuine public concern about environmental pollution and work was under way to increase efficiency and reduce emissions of sulphur and nitrogen oxides, it was also the time when there were being developed, nuclear power plants with no emissions, and the combined cycle with high efficiency and much greater operational flexibility.

Many coal-fired plants built around that time are still operating and while in Europe some of these must be shut down by 2015, those that remain are probably being retained to ensure that there is a secure electricity supply until new, more environmentally friendly capacity can be installed. Because why else would anyone fit FGD to a thirty-year old coal-fired power station?

No other government has followed Ontario in calling for coal to be taken out of power generation altogether. But then there are no coal mines in Ontario with all that it entails in closing down an industry. Ontario has moved the coal closures back five years to 2015 but is calling for a reduction of carbon dioxide emissions by up to two thirds during the remaining time. Limited operating hours and some biomass co-firing would help to achieve this, but it is clear that the four coal-fired plants will have to continue in operation until new nuclear or gas-fired plant can replace them.

In January 2009 Progress Energy, of Raleigh, NC, awarded a contract to Westinghouse and Shaw Group for a new nuclear plant in Levy

County, Florida which will have two 1105 MW Advanced Pressurized Water Reactors. These will cost $7.6 billion and are planned to be in operation between 2016 and 2018. On completion the company intends to shut down 866 MW of coal-fired capacity at its Crystal River site.

No other country or economic region has introduced an equivalent measure to the European LCPD which is not directed only at coal but rather at old technology in a large tightly integrated power system. It is a measure directed at more environmentally friendly technology of higher efficiency, lower emissions and lower environmental impact and less impact on public health.

Carbon capture and storage is an established techique in the oil industry for enhanced oil recovery and purification but with smaller volumes of gas. The experiments in power generation have only just started and all are based so far on the use of slipstreams yielding at most a million tons per year of carbon dioxide. We may not see any great use of carbon capture until after 2020

It is becoming increasingly difficult to get consent for new coal-fired power plants in the United States and parts of Europe. It all stems from uncertainty over carbon capture. If it is a matter of what is the best available control technology for carbon dioxide emission reduction then it is ultra supercritical steam with its higher efficiency.

The Green lobby are still banging on about global warming and "saving the planet" but is this really another front to their traditional anti-nuclear, pro-renewables stance. We cannot take them seriously if on the one hand they argue that we reduce our energy consumption but then on account of their own irrational fears of the alternative, advocate a concept of clean energy production which will create so many more energy loads, because of its lower efficiency, and push up the price of electricity, thereby adding to inflation.

If we look at some of the advanced industrial countries of the Far East which import about 80% of their energy resources, they are closer to the goal of taking coal out of power generation. Japan and Korea, particularly aim for 60% of nuclear power supply and in recent years almost all of their other new power plants have been gas-fired combined cycles, with several of the Korean plants designed as combined heat and power systems for district heating in new town developments.

Therefore the inevitable conclusion is that it must be good bye to coal as a fuel for power generation. We will need it for the manufacture of clean fuels and other chemicals when oil has become too expensive and shows signs of running out. Then nuclear generated power and heat will be needed to provide the process energy.

5

A nuclear energy revival

Nuclear power is the least understood by the general public, the most complex to implement, the most emotionally opposed and yet potentially, the most beneficial of all the energy systems. It operates without any greenhouse gas emissions and has the potential to produce more fuel than it consumes.

The understanding of the role of atoms as a source of energy dates from December 1942, with Enrico Fermi's Chicago pile which first demonstrated a controlled chain reaction.

Sixty-seven years later we have passed through a period of Green economic sabotage, there is no other way to describe it, in which nuclear power development was almost stopped in its tracks in North America and Europe where the Green movement and its anti-technology posturing had taken hold. At least five reactors in Europe have been shut down for blatantly political reasons.

But now there is the beginning of a revival, not only in the building of new versions of the existing water-cooled reactors, with the first two in Europe under construction since the completion of the last, Temelin Unit 2, in the Czech Republic, in 2003. In the United States one unit is under construction, the second for Tennessee Valley Authority's Watts Bar site, where Unit 1 was the last reactor to go into service in the country in 1996.

On the other side of the world the scene is markedly different. The four economic powers of the region, China, Japan, Korea and Taiwan, have continued to develop nuclear power for specific reasons. China needs to meet rapid economic growth and combat severe air pollution. It has 11 operating reactors with a total capacity of 8587 MW and another 21 under construction which will add a further 22 000 MW by 2015.

Japan, Korea, and Taiwan have few indigenous energy resources and import almost all of their fuel. Japan despite its traumatic introduction to nuclear energy in August 1945, bought a 160 MW Magnox reactor from

5.1: Ling Ao, China: first two of four 1000 MW PWR supplied by Areva and completed 2004. Two more units are under construction. (Photo Courtesy of Areva NP)

the UK which went into operation in 1966 and ran for 32 years. They now have 55 operating reactors producing 47580 MW with another nine under construction which will add 12 262 MW by the end of 2017 at which time nuclear power plants will account for 40% of national electricity supply.

Korea currently has twenty operating reactors, with a total capacity of 17 533 MW, and nine under construction to add 9400 MW by the end of 2016. Long term the Government aims to have 60% of electricity supply from nuclear power by 2035.

Taiwan with six operating reactors and two more under construction has the 4884 MW of operating plant represeting 19.3% of Taiwan Power's capacity, but providing 25% of the national electricity supply because of its lower operating costs. Completion of the Lungmen plant on the northeast coast with two of the new GE Advanced Boiling Water reactors will add a further 1300 MW each in 2009 and 2010.

These economic powers of the Far East with their exports of cars, cameras, televisions, computers and other electronic goods around the world, and except for China, almost totally dependent on fuel imports, have completely embraced nuclear power in sharp contrast to the attitudes in the countries of the European Union and North America who invented the technology, and some of which are still openly hostile to it.

A few Governments have now seen the light, and are planning to

5.2: Biblis,Germany; General Election in 2009 could decide whether these and two other reactors will be closed for political reasons. (Photo courtesy RWE Energie)

install their first nuclear power plants for more than twenty years. Meanwhile, sustained by a few export orders to the Far East, and service business on the existing plants, the industry has been quietly designing new reactors which are simpler to construct and of higher performance. Eleven have been ordered including six for three sites in China, and three already in operation in Japan.

To understand how nuclear energy, after a confident beginning at the end of the 1950's, was reduced to the point that for fifteen years after 1990 only five reactors started in operation in Europe and two in the United States, and with many European Governments adopting openly anti-nuclear energy policies, we must look back to the origins of the technology more than 60 year ago.

Fermi's demonstration came at a time when the United States had been at war on two fronts for a nearly a year. There was enough knowledge of nuclear physics in Europe that a British commando raid had been made on German-occupied Norway to destroy a heavy water plant at Telemark, lest the the Germans were themselves on the route to making the first atomic bomb.

Meanwhile, in July 1945, the demonstration of the first atomic explosion at Alamagordo, New Mexico, was of itself such a terribly awe-inspiring event that the subsequent bombing of Hiroshima and Nagasaki were seen by everybody in a different light. Nothing like it in its destructive effect had been seen before, not even the "conventional" fire bombing of Dresden, Germany, in February 1945, nor thankfully ever since. No nuclear weapons have been used since but the threat to use them effectively kept the peace in Europe for more than 60 years.

Nuclear bombs may have suddenly ended the Second World War, but the bombing also created the opportunity to study among the survivors, many of whom were still alive in 2000, the long term effects of radiation on the human body. Nuclear weapons would never again be used, but the association of nuclear power for which the fuel was the same material as was used in the bombs is still in the minds of the most adamant opponents of the technology.

It is on the basis of this knowledge that the nuclear industry has evolved a safety culture and a care for the health of its workforce which is unparalleled in any other industry. It is probably true that more people have died of cancer after a lifetime of work in a deep coal mine than have done so after a similar time working in a nuclear power plant or any of its supporting services.

Also because of the radioactivity of, particularly, the spent fuel, the nuclear operators have been held responsible for the safe disposal for all time of all their waste materials. No other electricity generator has had this requirement placed upon it, until the Greens thought up carbon capture and sequestration from coal-fired plants to combat global warming.

There have been basically four different types of reactor developed for power generation, but the method of operation is common to all. Fuel rods, which can be of either natural or enriched uranium, depending on the type of reactor, are assembled into bundles and placed in a tank holding a moderator to control the rate of reaction and with a coolant passing over the fuel bundles to remove the heat to an external steam generator.

Water is both moderator and coolant in the water-cooled reactors. Heavy water is both moderator and primary coolant for the heavy-water reactors, and graphite is the moderator with either carbon dioxide or helium coolant in the gas-cooled reactors. The Russian-designed RBMK reactor has a graphite moderator and water cooling.

Although the nuclear reaction is not a combustion process but an atomic reaction in the solid state, the industry has adopted the combustion analogy in referring to the rate of consumption of the fuel as the burn-up rate.

Control rods, containing Boron as a neutron absorber, are used to control the rate of the reaction. With all the rods in the core the reactor stops and as they are withdrawn so the reaction increases up to the full thermal design when the heat produced in the reactor and the heat removed as steam are in equilibrium.

The uranium fuel cycle is based on the isotope U_{235} which occurs naturally in uranium ore at about 1%. For a nuclear reactor this isotope must be concentrated to at least about 3.5%, although the original Candian-designed CANDU reactor and the British Magnox gas-cooled reactor use natural uranium. The new Pebble Bed Modular Reactor (PBMR) under development in South Africa will have fuel enriched to 10%. which contributes to its higher efficiency and the fact that the fuel pebbles could stay in the reactor for up to six years.

Enrichment starts with converting the uranium oxide to uranium hexafluoride, which is a gas above $57°C$, making it ideally suitable for the enrichment process. The original diffusion process, was highly energy intensive given the difference in atomic weight of the two uranium isotopes is about 1.3%. By 1970 the centrifuge was found to be more efficient and require less energy. It also used a large number of high speed centrifuges connected in series.

This is the basic fuel preparation: enriched uranium, converted to uranium oxide is made into fuel elements and the depleted uranium stripped of the radioactive isotope is available for the next step of the fuel cycle, or its properties can be exploited for industrial use.

In the reactor the U_{235} atoms lose neutrons and break up into isotopes of other elements of less than half the atomic weight of uranium, and the energy which held them together is released as heat. Faster neutrons also collide with the U_{238} making up 97% of the fuel element to produce plutonium. This is one of a series of transuranic elements above uranium in the periodic table, which is also fissile, so that some of its atoms similarly break up and add to the heat output. Reactors with unenriched natural uranium produce fewer fission products and more plutonium.

Eventually, after about a year in the reactor, some of the original fuel elements have "burnt up" and must be removed. Since the core is of circular cross section, the reaction is stronger in the centre than at the outer diameter, so not all the fuel elements burn up at the same rate. About a third are taken out every year and replaced with new elements. What remains is shuffled to new core positions to improve the rate of burn up.

The spent fuel elements consist mainly of unconverted U_{238} but with a quantity of plutonium (Pu_{241}) and fission products, mainly strontium (Sr_{90}), caesium (Cs_{136}), and iodine (I_{131}). All of these are radioactive isotopes with half lives ranging from a few days in the case of I_{131} to about 24000 years in the case of Pu_{241}.

The next step of the fuel cycle is to reprocess the spent fuel from the reactor. This is basic chemistry which separates the uranium, and plutonium from the fission products. The actual nuclear waste is the fission products which since they are derived from the 3.5% enrichment of the original fuel elements cannot be more than 35 kg for every ton of fuel. A 1000 MW pressurized water reactor would hold about 100 t of fuel, about one third of which is changed every year. So potentially every year 33 tons of spent fuel is sent to reprocessing and of this about 990 kg is high level radioactive waste, amounting in total to about 60 tons over the 60-year life of the reactor.

To complete the fuel cycle the plutonium and uranium as their oxides can be made into fuel elements and returned to the reactor. There are two ways that this can be done. First the plutonium oxide can be mixed with enriched uranium oxide to produce a mixed oxide fuel for the same reactor. The uranium and plutonium isotopes will split and create more plutonium which can again be separated from this spent fuel charge to be mixed with more uranium oxide for another fuel charge.

This is the method in current use in a number of countries. Recycling in this way extends the original uranium resource so that more of it is used and in practice the fuel can be reprocessed several times until all of it has been converted to plutonium, and the maximum amount of energy obtained from its fission. Many reactors in Europe and the Far East are running on mixed oxide fuel, but only in Japan is it a national energy policy that spent fuel will be reprocessed and the recovered plutonium used in mixed oxide fuel for the existing reactors.

All nuclear power reactors to date use a uranium fuel cycle, but this is not the only one possible. Thorium is a more stable element than uranium and is considerably more abundant. There are particularly large deposits in southeastern India.

Thorium is not naturally radioactive but can absorb neutrons to breed another uranium isotope, U_{233}, which is also fissile. Therefore a thorium fuel would be a mixed oxide assembly with plutonium oxide to produce the neutrons to kick start the reaction.

The new advanced CANDU reactor, the ACR 1000 is designed to work with the Thorium fuel cycle as are the CANDU derivative rectors being built in India. The original aim for the commercial design of the German Pebble Bed Reactor some thirty years ago was to use a Thorium fuel cycle. The versions currently under development in South Africa and China are initially using a uranium fuel, but could switch to Thorium in later versions..

In Russia the aim is to use thorium fuel seeded with plutonium in the

VVER 1000, the PWR on which they are extending their nuclear power program. The plutonium seed for this fuel is recovered weapons-grade material.

The other way is to move to a plutonium fuel cycle with a fast breeder reactor. This is a reactor of higher power density which has a plutonium core surrounded by a blanket of natural or depleted uranium. As the plutonium core burns, fast neutrons cause more to be created from the uranium in the blanket. The spent core and blanket materials are reprocessed and new material added to make a fresh core and blanket. Of three commercially sized fast breeder reactors in the UK, France, and Russia, only the Russian unit, rated 600 MW, at Beloyarsk is still running after 28 years.

These are the basic processes of the nuclear fuel cycle from which it can be seen that only a small part of the fuel charge is actually consumed at any one time. The reaction produces plutonium which also contributes to the output and which effectively exploits more of the energy potential of the original fuel charge.

If however, reprocessing is not carried out, then the spent fuel must be included in the high level radioactive wastes, and would amount to over 22,000 tons over the 60 year life of the reactor. With a fleet of 100 reactors all operating on a once through fuel cycle there would be over two million tons of high level waste yet of considerable energy potential, to be prepared for storage and deposited in a deep underground repository where it would be inaccessible for all time.

If nuclear technology had not developed against the background of the Second World War, the situation today might be a lot different. The gas-cooled reactor might be the dominant type rather than the water cooled designs which were born out of the competition to develop a power plant for a submarine. There might never have been a campaign for nuclear disarmament to colour opinions. The technology might now be more widely used and more willingly accepted by the general public. But this was not to be.

The years immediately after the Second World War were seen as a new dawn, as Europe and Japan set about reconstruction. The United Nations had replaced the discredited prewar League of Nations, and while the scientific community had agonized over the American monopoly of the new weapon, first the Soviet Union, then the United Kingdom, and later, France and China developed their own atomic bomb and for a long time afterwards these five were the only countries to have nuclear weapons.

In November 1951, Dwight D Eisenhower, the former commander

in Europe of the Allied armies at the end of the Second World War, was elected President of the United States. The war in Europe had been over for six years and although other countries had demonstrated that they had the bomb, Eisenhower saw that if nuclear energy was controllable, as indeed Fermi's original Chicago pile had shown it to be, then it would be possible to build a nuclear reactor to produce steam to generate electricity. This was the peaceful use of atomic energy which was to became a central policy of the Eisenhower administration.

The Nuclear Non-Proliferation Treaty was devised as a means of sharing nuclear power technology without encouraging the spread of nuclear weapons. Those countries which signed up to it undertook not to develop nuclear weapons in return for which they could share the nuclear power technology and benefit from its use. The initial signatories were mainly in Western Europe and Japan and all of these countries had strong economic links to the United States.

This was a time of experiment to find a reactor system which would best be able to supply energy for electricity generation and did not have any underlying problems which could give rise to instabilities that could destroy the reactor. A Swedish pressurized heavy water reactor and the British steam generating heavy water reactor were casualties of this process.

Basically three types emerged: gas-cooled, water-cooled, and heavy-water-cooled. Of these, it is the water-cooled designs, the Pressurized Water Reactor (PWR) and Boiling Water Reactor (BWR) which have come to dominate the market and account for the majority of the 454 nuclear power reactors that have been built to date.

At the time they were first developed, the strong economic power of the United States which had helped the post-war recovery of Europe with the Marshall Aid program, had also strengthened economic ties with those companies who had been their licensees pre-war; among them Alstom in France, AEG in Germany, and BTH in the UK. It was natural that if any one of these companies wanted to acquire nuclear technology and they were linked to an American company engaged in it, they would look there first. But although the Westinghouse PWR and the GE-designed BWR account for the majority of nuclear power plants in service today, some of the earliest designs were for gas-cooled reactors.

The first commercial nuclear power stations were at Calder Hall in the UK, with four gas-cooled reactors, which was opened by Queen Elizabeth II in September 1957; Shippingport, PA, in the United States, with the first PWR, followed a year later and then Ågesta, in Sweden, which was a small heavy water reactor installed in a cave in the southern suburbs of Stockholm powering a steam turbine connected to the local district heating network.

As these first nuclear plants came into operation there was great public euphoria in the fact that they had been successfully built and that given the power of the atomic reaction, could it not be potentially an unlimited energy supply so that electricity could almost be given away.

The reality was somewhat different. Although the early plants performed well without any problems they proved not to be the cheap source of energy that people had imagined. It must be remembered that by 1960 the largest generators on the European grid were about 200 MW and the majority were much smaller.

The four reactors at Calder Hall each supplied steam to a 50 MW turbine which provided both power and process steam to the neighbouring British Nuclear Fuels Ltd, Sellafield works. In Sweden, the Ågesta reactor was rated 10 MW and supplied 65 MJ/s to the district heating network of Farsta, a suburb of Stockholm. It was shut down in 1974. The prototype PWR at Shippingport, was rated 68 MW and ran for twenty-five years.

So nuclear energy began in Europe as combined heat and power. But no more nuclear combined heat and power plants were built, until Gösgen Däniken, in Switzerland which came into operation in November 1979, and then for a particular environmental reason. The site was in a valley and to take process steam off the turbine to supply to a neighboring pulp and paper mill would reduce the condensing tail and lower the risk of the cooling tower causing fog at certain times of year.

Separately, in what is now Slovakia, a distirct heating network was installed in the neighbouring town of Trnava from the Bohuniche nucleat power plant which went into operation in 1987. Ten years later the network was extended to two other towns in the area.

In practice, outside of the nuclear weapon states, only Sweden and Canada successfully developed their own reactor systems, but in France and Germany small PWR and BWR reactors were installed from which their industries have developed 1000 MW and larger units which have been sold around the world

Development of gas-cooled reactors started in the United Kingdom and France but it is only in the UK that development was continued with the Advanced Gas-cooled Reactors, of which fourteen were built each rated 660 MW with two reactors on each of seven sites. These and the earlier Magnox reactors were cooled by carbon dioxide. Other gas-cooled reactors are helium-cooled.

One of the British Magnox reactors was sold to Japan and another to Italy, and France sold one of their gas-cooled reactors to Spain, but apart from these, no other gas-cooled reactors have been sold outside the countries that developed them.

Canada developed CANDU a natural uranium, heavy-water-cooled and moderated reactor with a unique tubular core structure separating the two heavy water streams, and with a heavy-water/light-water heat exchanger supplying the steam turbine. CANDU systems have been sold to India, Pakistan, Argentina, Korea, China, and Romania.

The first nuclear power plants had appeared on the scene at a time when there was considerable development taking place in the production of larger steam turbines and generators and the introduction of higher transmission voltages. The first high voltage direct current (HVDC) connexions were laid in Scandinavia and across the Dover Straits between France and England.

The effect of these developments was to reduce the cost of generating and transmitting electricity through the economies of scale in a familiar technology. For the utility management, introduction of nuclear power was a new technology with special requirements in training, operation, maintenance, fuel management and safety and it would need to have a strong competitive advantage before they would consider it.

Fifty years later as the price of oil and the other fossil fuels started to rise, the high energy density and relatively low cost of the uranium fuel has meant that nuclear generating costs were not rising as fast as those of the fossil-fired power plants.

The American water-cooled reactor designs were born out of a competition to develop a nuclear power plant for a submarine. Quite apart from the purpose of the submarine, the fact that a nuclear reaction is not a combustion process makes it an eminently suitable power source for the boat, since it would stop it having to surface to recharge batteries and could remain submerged for several weeks at a time.

All four nuclear weapons states developed nuclear submarines, but the Soviet Union also developed a nuclear icebreaker. In 1953 it was decided to develop a fleet of nuclear-powered icebreakers to develop the northern coastal areas of Russia and Siberia. This fleet of icebreakers would keep Arctic sea lanes open to supply the scientific and mining communities up there. The first nuclear powered icebreaker, *Lenin* was constructed at the Admiralty shipyard in St Petersburg and went into service in 1969.

Over the next thirty years six more icebreakers were built each with two 30 MW reactors for propulsion and electicity supply. Two smaller vessels each with a single reactor were built at the Wartsila yards in Finland, and delivered in 1989 and 1990. The two oldest icebreakers, *Lenin* and *Sibir* have been decommissioned and and a replacement vessel went into service in 2007.

The Arctic Fleet is an important commercial link along the North coast. The ice breakers have a cruising capacity of 6 to 8 months and and the reactors are refuelled every five years, The present fleet has seven icebreakers and five supporting vessels including *Sevmorput,* which is said to be the world's first nuclear-powered container ship. The fleet is owned by a division of Rosatom and operated by Atomflot out of their base in Murmansk.

Another shipping application was to the aircraft carrier which again allows it to stay out on duty for longer, since all the liquid fuel that it carries is then for its aircraft. The reactor supplies all of the electric power, heat and propulsion for the vessel.

Surface ships for general cargo have not been successful. In 1962 in the United States the nuclear cargo ship *Savannah* was launched under the Atoms for Peace programme but it came at a time when much larger oil-fueled cargo ships were being developed to carry containers as well as the super tankers for bulk oil transport.

Savannah proved to be an expensive way to carry cargo, not least because of the nuclear engineers that it had to carry in addition to the crew to operate the reactor, and was heavily dependent on the Federal subsidy that it received as a unique ship. Nevertheless it operated for seventeen years and was retired in 1979. The spent fuel assemblies that had been produced were later sent to France for reprocessing and fabrication into mixed oxide fuel which was supplied to one of the Tennessee Valley Authority's reactors.

In 1968 in Germany the nuclear powered cargo ship *Otto Hahn* was launched at a shipyard in Hamburg. The ship contained a small pressurized water reactor which had a thermal output of 38 MJ/s driving a steam turbine developing 10 000 shp. For the next eleven years the *Otto Hahn* covered 100 000 km on some 120 cargo carrying and research voyages, but as with the American ship, the running costs could not be covered by its commercial operations alone.

No other commercial ships were built and nuclear energy has been confined to the production of electricity and process heat. The only nuclear ships on the high seas are the submarines and aircraft carriers and the Russian ice-breakers.

The late 1960's were a time of cheap energy. Oil was only $3.00/bbl and even with the added taxes, gasoline in the UK was the equivalent of 6p/litre and the domestic rate for electricity was 0.5p/kWh. A further factor in the equation was the political interests of governments and their supporters. At the end of 1973 there were 98 reactors in commercial operation in fifteen countries and as many more under construction.

The Middle East war and the oil crisis flowing from it brought a sea change to electricity supply. Environmental concern for acid rain was beginning to alter the economics of coal-fired generation. The first combined cycles had appeared and bigger gas turbines of 90 to 150 MW, and running at synchronous speed, were under development. New gas fields were coming into commercial operation around the world. Nuclear economics were marginal, but could be improved by going to larger unit sizes, first 900 MW then 1200 MW and ultimately 1400 MW, in France.

Of all the generating systems nuclear is the only one that uses a fabricated fuel, and because of the radioactivity of the spent fuel the industry is legally required to store its wastes in a safe environment away from human contact. This puts a price on the fuel and on the power plant itself when decommissioned at some time in the future.

A nuclear power plant coming into operation in 2008 would not be shut down until 2068, by which time the people who ordered, built and commissioned it, and the first operating staff of the station would almost all be dead. Back in 1973 nobody could say with certainty what would be the operating life of a nuclear power plant. Capital amortization was set at 25 years because that was what was done for the other power plants. Now many nuclear plants have been granted life extensions to sixty years.

Also in 1973 the first nuclear stations had not produced such a large volume of spent fuel that they could not store it in ponds at the power station. If it was kept there for a few years many of the short lived fission products would have gone through at least one half life and possibly several so that they would be less radioactive when it came to reprocessing the spent fuel.

Radioactive waste is classed as low, medium and high level, and does not all come from nuclear power stations. A considerable amount of radioactive isotopes are used in the X-ray departments of hospitals and of course there is military weaponry which deteriorates with lack of use over time and must be scrapped.

All this material is potentially a source of energy. It must be be first reprocessed to separate the uranium and plutonium from which mixed oxide fuel is placed in a reactor to generate more electricity. Some of this is being done with military stocks released as the result of the Strategic

Arms Limitation talks between the United States and the Soviet Union at the end of the 1980's.

With so much going for it as a clean and increasingly economic energy system, and with fuel supply stretching hundreds of years into the future, how did nuclear energy get into the present state where in much of Europe, it is regarded as an energy of last resort and where governments of three countries with large nuclear power supplies have planned to take nuclear energy out of electricity generation. We must look back to the oil crisis of 1973.

For more than 20 years there had been an active campaign for nuclear disarmament which had instilled in its supporters the belief that plutonium was the material of bombs and to use it as fuel for a power reactor was asking for trouble and laying one open to risk of the fuel being stolen by terrorists to make a bomb.

The Green Movement which was growing rapidly in North America and was beginning to move into Europe had started as a movement for environmental protection but very soon realised that using plutonium as reactor fuel was too dangerous; after all it had a half life of 24000 years. We didn't want to leave these dangerous plants around for future generations to clear up.

But for every kilogram of plutonium that is made into fuel pins for a nuclear reactor to generate electricity, there is one less kilogram with which to tempt the terrorists, who in any case have had far greater success in Madrid and London with simple chemistry and using easily obtainable materials in everyday use

The first big challenge to nuclear power came when India detonated a nuclear device in the Rajasthan desert in 1974. Even though many countries had signed up to the Nuclear Non-Proliferation Treaty, there was the nagging fear that some countries would use their nuclear knowledge to make a bomb. Canada, having supplied a 220 MW CANDU reactor, to India, which had gone into service in December 1973, immediately stopped all nuclear cooperation, including the supply of equipment for a second plant being built alongside it.

India had then a large number of nuclear engineers and already had two 160 MW Boiling Water Reactors at Tarapur, in Maharashtra State, which had gone into operation in October 1969. The country was determined to adopt nuclear power and had based their future plans on the CANDU system which at 220 MW could be placed around the country closer to the loads than many of the coal-fired and hydro resources, and this would also facilitate the spread of rural electrification. Despite Canada's withdrawal, India has two BWR and seven CANDU derivative reactors in operation with a total capacity of 1720 MW, and has continued development of a 500 MW version of the CANDU reactor.

Thirty five years have passed and India has signed nuclear cooperation agreements with the United States and in February 2009 Areva won a contract to supply two of their 1600 MW EPR reactors with an option for four more. Nuclear power at present accounts for 3% of total power ouput in India, but will increase with the completion of four more locally-built CANDU derivatives,

In Europe the first threat to nuclear power development came with the election of a Centre-Right Government in Sweden in September 1976. It was the first change of government for more than forty years. It inherited a nuclear power plan for twelve reactors which were all under construction, and there were plans for three more, one of which, at Osterhaninge, would supply district heating to Stockholm and two more in the south at Barseback would supply district heating to Malmö, Hälsingborg, Angelholm, and Lund. But the new Prime Minister, Thorbjorn Fälldin, was a well known anti-nuclear activist, being a northern sheep farmer under whose land had been found a large uranium deposit.

In September 1976 there were four nuclear plants in operation and two more in the final stages of construction and less than a year from commercial operation. Furthermore Fälldin was in a coalition with two strongly pro-nuclear parties so was restrained on the nuclear issue. But the outcome of Swedish deliberations over the following years was a Nuclear Law which first required all nuclear plants to set up contracts for reprocessing spent fuel, and second, called for the construction of a safe underground repository for the long term storage of nuclear waste.

In the United States at the 1978 mid-term elections in California, the anti-nuclear Proposition 13 was included in the ballot. It had received enough support to be included as a referendum issue, as American electoral law allows. The Proposition was that there should be no nuclear power in California, which already had two reactors in operation and three more under construction, at the two sites: Diabolo Canyon north of Los Angeles near St Louis Obispo, and San Onofre, on the coast midway between Los Angeles and San Diego. Although Proposition 13 was defeated, this did not stop the cancellation of a third project, Rancho Seco, near Sacramento.

Plutonium, which had been the fissile material of the Nagasaki bomb, was a reaction product of the reactor, and would be the fuel for the fast breeder reactor. Enough was known about it, that it was a transuranic element which was not naturally occuring, and had a very long half life, and that it was a mild alpha emitter.

It was enough information to strike fear into the general public if skillfully presented. We could not have a reactor with a plutonium fuel cycle where any country with reprocessing capability could take out some of the core material and make a bomb. The basic anti nuclear argument was that we should not have to leave future generations to clear up the mess that we had left behind.

It was a powerful argument at the time which mirrors what people feel about their own families. We all want our children and grandchildren to enjoy a better life than we have had, and one definition of that is that they should be able to depend on a secure supply of energy But, 60 years later, Hiroshima and Nagasaki are both thriving industrial cities exporting their products all over the world, which rather gives the lie to this basic Green argument.

As the Green Movement took hold in Europe, there arose great public concern over nuclear power in several countries. In Austria it reached the point that with the completion of the country's first nuclear power station at Zwentendorf, near Linz, the government decided to hold a referendum in the autumn of 1978 before fuel could be loaded and the plant brought into operation.

For a government in power a referendum is the only full test of public opinion outside a general election and the conventional wisdom is that voters will use the opportunity to protest against an unpopular government. So when Chancellor Bruno Kreisky announced that he would resign if the vote went against bringing Zwentendorf into operation, voters took him at his word and voted against nuclear power. The only problem was that Kreisky did not resign and the power plant was never put into operation.

Then in March 1979 one of the reactors at Three Mile Island, near Harrisburg, PA, tripped uncontrollably so that the reactor coolant was evaporated and the core started to melt. This was not known at the time but there was enough evidence that a major accident had occurred and that emergency procedures had to be brought into play. It being in the United States the event was played out in full view of the media and although there was no breach of the pressure vessel and containment and no release of radiation to the outside, this was nevertheless a golden opportunity for the anti nuclear brigade to restate its case. If it had happened once, couldn't it happen a second time elsewhere?

A week later in Canada, there was a more serious non-nuclear event which was not reported around the world. This was the Mississauga train crash, near Toronto, when a freight train derailed at a level crossing. There was no nuclear material on board but the two wagons which jumped the tracks were tankers carrying respectively liquid methane and liquid chlorine.

Chlorine was the gas weapon used in the first World War. There were plenty of people alive then who would have grown up conscious of relatives who had been gassed in that war and what its effects were. While nobody was evacuated from Harrisburg, half a million people were evacuated from around the Canadian crash site such was the fear that if any leakage from the methane tanker were to ignite it would cause an explosion which would release into the atmosphere from the adjacent car the several tons of chlorine that it carried.

Three Mile Island showed that a nuclear reactor could suffer a major loss of coolant and remain within its containment and have no impact on the environment. A week later the event in Canada showed that there are dangerous substances carried through big cities on the railways which, in the event of an accident can cause either serious loss of life, or great inconvenience, which are not related to anything nuclear, and are largely taken for granted.

But the damage had been done. The US President at the time, Jimmy Carter, was not openly anti-nuclear, having been an officer on a nuclear submarine, but he had come to power after the scandal of the second Nixon presidency, and undoubtedly had a large Green following among his Democratic party supporters.

There was a lot of argument at the time about the merits of reprocessing the spent fuel. Reprocessing plants had been built at Sellafield in the United Kingdom, and at Cap La Hague in France to handle fuel from their own gas-cooled reactors and had started to take orders from other nuclear operators in Sweden and Germany and Japan.

The first nuclear power plants had also been built in Japan, Korea and Taiwan, countries which had very few indigenous energy resources and had imported nuclear fuel services from Europe and North America. They quickly realised that if they imported uranium they should use as much of it as possible, which initially meant sending it to Europe for reprocessing and fabrication of mixed oxide fuel, and in the long term building their own fuel production and reprocessing facilities.

Then as delegates to the 1979 International Nuclear Power Conference assembled in Vienna, the announcement came from the United States that a planned reprocessing plant under construction at Barnwell, South Carolina, was cancelled and that no American nuclear operator could reprocess spent fuel. They would operate a once-through fuel cycle and the spent fuel would be classified as high level atomic waste and stored at the plants until such time as a deep underground repository could be constructed to receive it.

It seemed at the time that very few delegates could take the American position seriously. Operating licenses for their nuclear plants were set initially for 40 years and it seemed as if the United States Government had swallowed the Green anti-nuclear argument hook, line and sinker and would give up nuclear energy completely by the end of the century.

Asian and European operators were particularly angry becuse they came from cultures with a tradition of energy economy. For them it meant making the greatest use of the energy contained in the uranium fuel at a time when utilities were buying up several years uranium requirement from the various producers.

The compromise solution was the International Nuclear Fuel Cycle Evaluation, which would report back and if it was found that there was a sound economic case for reprocessing and recycling nuclear fuel then those countries that needed to do it could. Effectively the United States went its own way, although the three reactor producers continued to sell around the world, and particularly in the Far East which has been the only significant nuclear market in the first decade of the 21st century.

Spent fuel from European and Japanese reactors was shipped to the reprocessing plants in France and the UK, under the condition that the recycled material was returned as fuel elments and the fission products would be returned in a secure form for disposal at their own facilities: in Sweden at Forsmark on the Baltic coast, in Germany at Gorleben, and now in Japan at Rokkasho Mura.

By March 1980, in Sweden, Fälldin's government had been replaced by a minority government of Liberal and Conservative members who were in the main pro nuclear. The Swedish people had accepted nuclear energy as an indigenous technology and apart from a few small oil-fired plants powering district heating schemes in the major cities, electricity came from hydro plants on the major rivers, and six of a planned programme of twelve reactors in operation.

Three Mile Island had called nuclear power into question among the politicians and to settle the issue once and for all, the decision was taken to call a referendum on the future of nuclear energy. Swedish voters had two options: run the six reactors for a further ten years and abandon construction of the other six, of which two were complete and had been waiting two years for consent to load fuel; or complete the programme of twelve units and run them for their design life, put at about 25 years.

The referendum voted for continuation of the programme but the third question that should have been asked, and was not, was whether any more reactors should be built after the twelfth of the then current plan.

Then on April 26, 1986 the nuclear industry had almost recovered from the trauma of Three Mile Island and the American abandonment of the fuel cycle when the accident which would drive the Green anti-nuclear lobby into an ecstatic frenzy finally happened at Chernobyl, Ukraine. One of four reactors, on which a low power test was being performed, ran away causing a massive explosion which blew the top off the reactor building and spread a plume of radioactive particles northwards over Byelorus towards Scandinavia where two days later evidence was picked up on radiation monitors at the Forsmark nuclear power station north of Stockholm and later confirmed elsewhere.

It would be no exaggeration to say that the result of Chernobyl was the collapse of Communism. The hallmark of the Soviet Union until then had been their excessive secrecy of industrial and commercial affairs. If a western company sold them something they would be paid promptly for it, but that was as far as it went. Chernobyl was something different. It had four 1000 MW type RBMK reactors, which were unique to the Soviet Union, but the after effects had been felt in other countries. All nuclear operators had to know how this accident had happened and whether it could happen in any other type of reactor.

The Soviet industry had also kept a very tight rein on the fuel cycle. They had developed a 440 MW PWR which they had exported to Hungary, Bulgaria, Czechoslovakia, Cuba and Finland, all the fuel was supplied from the Soviet Union and spent fuel had to be returned to them for reprocessing. But the Soviets had also developed the RBMK a graphite moderated, water-cooled reactor which was considered to be primarily a plutonium producer. Chernobyl had four reactors of this type, but also four more had been installed at Sosnovy Bor, near St Petersburg, two at Beloyarsk and two 1500 MW units at Ignalina, Lithuania, where one unit was shut down at the end of 2003 and the other may be shut down in 2009.

The immediate action was to bring western experts in to look at the safety systems of the entire Soviet commercial reactor fleet. To close all the RBMK reactors would lead to a serious shortage of electricity, but their safety systems had to be modified to ensure that the accident couldn't happen again elsewhere. At the time there were thirty nine operating reactors in Russia and seven more in Ukraine, including the three undamaged units at Chernobyl.

By this time anti-nuclear opinion was gathering strength in the

Western political classes. The United States might have had a President who had served on a nuclear submarine, but nowhere in the European Union was there a Minister in any Government who had earlier been superintendent of any sort of power plant, let alone a nuclear power station, although the British Prime Minister at the time, Margaret Thatcher, had trained as an industrial chemist, and at least one of her ministers was a Chartered Engineer.

By 1995 nuclear power was no longer being considered in Western Europe at least. When the first nuclear power plants appeared nobody really knew how they would perform in the long term. They had a reasonable idea of the life of a generator and it was generally the case that the plant wouild be built in the expectation that it would run for twenty five years. So the nuclear plants were similarly licensed for 25 years, with the result that almost all of the European nuclear power plants could be shut down by 2015.

Since Chernobyl there had developed an underlying anti-nuclear feeling in a number of European governments. But only in Italy was there the complete closure of the country's three operating power plants and construction stopped on a fouth unit at Montalto di Castro, which was subsequently turned into a combined cycle.

There were three countries each with a large nuclear components of electricity supply, Belgium, Germany and Sweden which passed measures in their parliaments to build no more nuclear power plants and shut down all of the existing reactors at the end of their design life. However, in Belgium, where seven reactors generate 65% of the electricity supply, the Government came to realise that they could not do without them and in 2006 decided to extend the operational life of all seven reactors for a further twenty years.

In February 2009 the Swedish Government announced that it would build more nuclear plants to avoid generating more greenhouse gases. Given that more than 90% of their electricity supply comes from nuclear and hydro power, this was the only way it could be done. The four existing sites at Forsmark, Oskarshamn, Ringhals and Barseback would have their reactors replaced with new units at the end of their lives.

The strongest anti-nuclear protest was in Germany. During the Cold War this had been the front line between the armies of NATO and the Warsaw Pact. The country was divided by a fortified fence between the German Democratic Republic to the east and the Federal Republic of Germany to the west. The city of Berlin, which had been the capital of the united Germany up to the end of the Second World War, was similarly divided with the western suburbs of Berlin surrounded by a

5.3: Loviisa, Finland: These two Russian-designed 440 MW PWR have been built with Westinghouse containment, control and safety sytems. (Photo courtesy of Fortum)

high concrete wall which separated them from East Berlin which was the capital of the so-called German Democratic Republic.

Many Germans were from families that the barriers had divided and some had got out of East Germany before they went up. As the front line between the two superpowers, both nuclear weapon states, and, after the experience of the Nazi dictatorship from 1933 to 1945, which had brought Germany's situation about as the result of its defeat in the Second World War, it was inevitably going to colour opinions. It is no surprise that Green protest developed first against nuclear power in Germany.

From 1997 Germany had a Centre Left coalition government of Social Democrats and the Green Party, who wanted to shut down all the nuclear stations and erect a forest of 2 MW wind generators to replace them. The nuclear power stations are still running and with alongside them a large number of wind farms, and in the east new, supercritical lignite-fired steam plants.

The anti-nuclear feeling had also infiltrated the council of ministers of the European Union. At this time they were seeking to bring into the Union the eastern countries that had only recently given up communism and which, in the case of the Czech Republic, Hungary, Slovakia, and Bulgaria had, between them a large fleet of Russian-designed 440 MW PWR's, and in the Czech Republic construction had started at Temelin, of the first of two 1000 MW PWR which were completed in 2003, after

5.4: Civaux, France. EdF's station was the first with the 1400 MW Type N4 reactor from which has been developed the EPR. (Photo courtesy of Areva NP)

they had joined the European Union. It was even suggested that these countries might not be allowed to enter the European Union unless they shut down their nuclear power stations.

Ukraine though not considering membership was similarly pressured to shut down the three surviving reactors at Chernobyl which had been extensively modified to western safety criteria. It even got as far as some of the western gas turbine companies proposing combined cycle schemes to replace it. Lithuania similarly was pressured because of the two RBMK reactors at Ignalina.

There are in 2008 some 439 nuclear power plants operating around the world, the majority brought into service before 1990. A few of the earliest dating from the 1960's have already been closed. The rest, however continue to operate, producing electricity without emitting one gram of any greenhouse gas.

Not only that but as base load plants many of them run continuously between refuelling and maintenance outages which are typically from 12 to 15 months apart. Where a nuclear plant trips, it is more likely to be due to an external factor such as a grid fault or an earthquake as happened in 2008 in Japan. But however it happens, safety systems come into play to shut it down.

There are five main types of nuclear reactor generating power today. The PWR is water-cooled and moderated and uses enriched uranium. It was originally developed by Westinghouse in the United States and licensed to Siemens, Framatome, and FIAT in Europe and Mitsubishi in Japan.

The Soviet Union developed a PWR which was produced initially as a 440 MW unit which was sold to countries within their sphere of influence in eastern Europe, and Cuba. Much was learned about the Soviet system because Finland bought two units for their Loviisa power station, but fitted them out with western instrumentation and control systems and a Westinghouse ice containment system. The first went into operation in May 1977 and the second in December 1980. The two reactors have operated reliably ever since, and have consistently been near the top of published performance tables.

The European reactor vendors developed their own version of PWR for the 50 Hz electrical system. In France, Electricité de France standardised on a nominally 900 MW design for thirty three units at nine sites. Two of six units on the Channel coast at Gravelines, midway between Calais and Dunkerque send their output over the HVDC link to England where EdF has a large customer base as the owner of London Electricity and Southeastern Electricity. The 900 MW units in France have been followed by a 1300 MW design which was later stretched to 1400 MW in the latest units to be installed. France has also met with some success in export markets and sold plants to South Africa, Korea and China in a nominally 1000 MW design.

Siemens developed the PWR as a 1200 MW unit in a spherical containment structure designed to withstand an aircraft crashing into it, It would be almost thirty years before such an event would take on a wholly new meaning in New York, but in those innocent times an air crash would be judged to be a small low-flying civilian or military aircraft getting into difficulties in bad weather and straying off course. Siemens has also exported plants to Spain and Brazil; all are 1200 MW class PWR supplied as turnkeys with their own turbomachinery and contol systems.

The other light water reactor system is the Boiling Water Reactor which was developed independently by GE in the United States and ASEA Atom in Sweden. The Swedish design introduced a number of innovations in reactor design which have been incorporated by GE in the Advanced BWR design of which the first examples were installed in Japan as the last two of seven BWR at Tokyo Electic Power Company's Kashiwazaki Kariwa station

The Swedish BWR has only been installed in Sweden and Finland and its development was conducted against the political background of anti-nuclear governments unable to convince a well informed public opinion which could see the advantages of nuclear power in terms of its environmental friendliness. The fact is that with more than 95% of electricity supply being provided in equal measure by nuclear and hydro power, Sweden has one of the cleanest urban environments in Europe.

GE's design has been hampered by the extended licensing procedures in its home market. One notable case is of two similar plants ordered at the same time, one for Taiwan Power and one for a US utility, the plant in Taiwan was already up and running before ground had been broken at the American site. GE granted a license to AEG, in Germany, later taken over by Siemens, and to Alsthom, in France, who completed the Caorso power plant near Milan, in northern Italy before pulling out in the reorganization of French industry 1975 to standardise on the PWR.

GE's big success has, however been in Japan where Hitachi and Toshiba have collaborated with them in the construction of 16 reactors for Tokyo Electric Power Company at Fukushima and Kashiwazaki Kariwa.

Canada developed its own unique system which comes from their being a uranium producer and having the capability to produce heavy water. The CANDU design is for a heavy-water-cooled and moderated system which consists of a horizontally mounted cylindrical tank, the Calandria, which is filled with heavy water and has parallel tubes running through it from end to end. The fuel elements are contained in the coolant tubes which can be accessed from either end of the calandria so that fuel pins can be shuffled on load to optimise the burn-up rate. CANDU plants of a standard 650 MW design have been exported to Argentina, Korea, China. and Romania, and two have been installed in Canada.

These are the current reactor systems which supply about 369 GWh. There have been few problems with any of the operating reactors, and only two serious accidents of which only the one at Chernobyl destroyed the reactor and released radiation into the environment.

Twenty years later, the nuclear revival is now firming up and it can be said that it started in the immediate aftermath of the Kyoto follow-up conference in 2001 with the Finnish Government's decision to build another nuclear plant, as the only sure way to meet its emission targets. It is as though the nuclear industry suddenly collectively realized that they had the technology to "save the planet" and it was a case of convincing governments of the strength of their argument.

The other post Kyoto reaction has been the irrational hatred voiced around the world of President George W Bush, as leader of the one nation that had not signed up to the Kyoto Protocol.

In fact what has been happening in the United States is that the Bush Government saw that nuclear power had to continue, and moreover that the fuel cycle should be taken up again since the closure forced on the industry by President Carter in 1979. By now all 104 operating reactors have various quantities of spent fuel in storage on their sites which represent a huge energy potential which it would be wrong to ignore, and the construction of the long-term repository at Yucca Mountain, NV is bogged down in the anti-nuclear posturing of the Nevada State Government

This revived interest in nuclear energy came at a time when fossil-fuel prices had risen sharply to unprecedented levels with oil passing $140/bbl in April 2008 although it had subsequently dropped back to $40/bbl by the end of the year. This has led governments to consider how they can control energy costs and have greater security of supply.

From their Vienna headquarters the International Atomic Energy Agency interpret the variously announced plans as a low forecast of 25% or a high of 93% in the increased contribution that nuclear energy will make to meeting the world's energy needs by 2030. Nuclear plants accounted for 15% of the total global capacity in 2008.

There have been two big actions of the Bush Government to restart the nuclear industry. First has been to revise the licensing system replacing the two stage licensing scheme of old with a single Construction and Operating License (COL) which is awarded for a certified reactor type and is granted for 20 years, after which it lapses if nothing is done. Some 32 New Start licenses have been applied for and so far twelve have been issued. The majority are for additional reactors to go alongside existing sites.

Second was the setting up of the Global Nuclear Energy Partnership (GNEP), the purpose of which is to close the fuel cycle in a secure manner so that the plutonium can be separated from spent fuel and made into mixed oxide fuel for the present generation of reactors around the world.

Effectively this plan recognizes that plutonium is an important fuel material. As soon as the reactor starts up it starts to produce plutonium which fissions and accounts for a significant part of the energy produced.

Besides mixed oxide-fuel production, GNEP will look at the design of the fast breeder reactors so that plutonium can be securely used as fuel for these reactors. In particular there is the Very High Temperature Reactor which with its high rate of burn-up would dispose of plutonium and the other transuranics to produce electricity and process heat.

The high price of fossil fuels since 2007 has led several countries to look hard at nuclear power and in more settled times the number of countries likely to be opposed by the Green Activists within government has somewhat diminished.

The other development is that the United States and others, having seen at first hand the benefits of nuclear energy with no emissions to damage public health, and low and stable fuel costs, have been asking why there cannot be smaller reactors produced which can give the same benefits to small countries and island networks which cannot use a reactor of 1000 MW.

This view is not new. The original developers of the Pebble Bed gas-cooled reactor in Germany in the early 1970's saw this as the main reason for its development: to produce a reactor that was inherently safe and that could be operated on any network anywhere in the world. At that time, the world had been shaken by a four-fold increase in the price of a barrel of oil, and were seeking ways to take it out of power generation and replace it with other fuels.

Forty years later it is still the relatively small networks of island utilities that have no access to gas and are still burning oil for power generation. It is only six years since gas supply came to the Isle of Man, and nine years since Singapore made a long term gas supply agreement with Indonesia which has since transformed their power system and greatly improved their efficiency of energy supply.

So there are now three small reactors under development which are aimed at this market of small electricity supply networks. The Pebble Bed Modular Reactor development in Germany was stopped in 1989 by an SDP/Green anti nuclear State Government in North Rhine Westfalia and was taken up by South Africa in 1990. There development of the reactor has continued a 160 MW prototype of the gas-cooled reactor with a direct closed-cycle gas turbine drive, is to be built alongside the existing nuclear power plant at Koeburg, north of Capetown.

Westinghouse have developed a 600 MW version of their APWR, but are also leading the development of IRIS the International Reactor Intrinsically Secure. This is a compact light water reactor rated at 300 MW which is being developed under GNEP. Westinghouse applied to the Nuclear Regulatory Commission for a type approval, which is expected in 2010.

The advantages of nuclear power are that it produces no emissions, and that the fuel costs are low. The high energy density of the fuel means that it accounts for a much smaller part of the total generating cost than does fuel for the fossil-fired plants. Now that the capital costs of many of the nuclear plants have been paid off, they have only operating costs and fuel to meet, with the result that the lowest generating costs in the United States are from the nuclear fleet.

So it is an economic issue for the world. How do we meet future energy demand and guarantee security of supply? How can we avoid building a power station and then finding three or four years down the line that the fuel price has rocketed and brought public protest at higher electricity prices? If nuclear power can avoid this and ensure that the lights stay on at a price we can afford, then it must be developed and there is really no alternative.

The small reactor developments are picking up the gas-cooled reactor technology again. The only operating gas-cooled reactors are now in the United Kingdom. But an economic small, intrinsically safe, high-temperature reactor would open up a market for industrial combined heat and power plants and district heating, which would sustain energy supply without any pollution or the need to strip carbon dioxide out of flue gases.

The intrinsic safety of the gas-cooled reactors stems from the fact that there is no possibility of a phase change in the coolant if the circulation fans failed. The control rods would drop to stop the reaction and the graphite core would absorb the decay heat and would not be damaged since the cooling gas is inert.

This is the extent to which the nuclear revival has progressed, largely unnoticed by the public at large. In the previous twenty years the hostility to nuclear power in much of Europe and North America has led to the contraction of the industry which largely survived on a few export contracts and some service business.

There also have been changes in the industry: Westinghouse Nuclear Business is now a division of Toshiba Corporation; Siemens nuclear business is now incorporated in Areva and have developed the 1600 MW EPR of which one each are under construction in Finland and France and two more in China. Westinghouse have also sold fourteen of their new design AP 1000 PWR, four of which are for two sites in China.

Three Asian countries with large nuclear power programmes, China, Japan and Korea, are also developing their nuclear construction industries. China has developed a 1000 MW design for a PWR in cooperation with Mitsubishi. Korea has standardized on the former Combustion Engineering System 80+ PWR and have built the first four at Yonggwang as OPR 1000 and have orders for eight more. System 80+ will no longer be built in the United States but is being developed as a standard 1400 MW design for the Korean market which will be assembled there and could eventually be exported.

So the future customer for a nuclear power plant has at least six new reactor designs to choose from and which have been certified in the United States or the European Union, although the Areva EPR design has applied for US certification which is expected to be awarded in 2009. These are all developed from existing PWR and BWR designs, but there are other smaller reactors under development in the United States, China, Japan and South Africa. If the prototypes are demonstrated and seen to work efficiently and to the predicted cost, then by 2025 we may see emerge a new group of industrial operators who will be running their first small nuclear plant.

Apart from the units in Finland and France no other reactors are under construction in Europe. In the United Kingdom the Nuclear Regulatory Authority is looking at three designs: the Westinghouse and Areva PWRs and the GE-Hitachi ESBWR. At the end of March the British and French Governments entered a nuclear technology cooperation agreement which suggests that the Areva EPR design has the upper hand, since EdF Energy, the British arm of Electricité de France, has a large customer base in London and south-east England, and now is the owner of British Energy, the nuclear operator.

Anti-nuclear protest in Europe seems to have calmed down perhaps because the Green movement seems to be trying to drive us up a dead end with clean coal and renewable technologies. Although the Green Party is out of power in Germany the present Government, led by the pro-nuclear Christian Democratic Union (CDU) has its hands tied by its partner the anti-nuclear SDP who had the Greens as their coalition partners in the previous government. Consequently an application by RWE Energy for a life extension for the first of two reactors at their Biblis station south of Frankfurt was initially rejected, and both these, and two other reactors will shut down in 2009 because they will have reached 32 years operation which was the limit set under the 2000 anti-nuclear law.

With now a number of units due to shut down, the present Govenrment has indicated that it willing to discuss extending licenses to 60 years as in the rest of the world, because it has realised the enormous cost that early closure would present to the German economy which is heavily dependent on manufactured exports. The result of the General Election of 2009 will determine whether this will happern or not.

New countries with nuclear ambitions are mainly in the Middle East: Turkey, Egypt, Bahrain, and the United Arab Emirates. Turkey apart,

TABLE 5.1: US REACTOR CERTIFICATION 2008

Design	Vendor	Output MW	Reactor Type	Certification Status
AP 600	Westinghouse	650	PWR	Certificated
AP 1000	Westinghouse	1100	PWR	Certificated
ABWR	GE	1370	BWR	Certificated
System 80+	Westinghouse	1300	PWR	Certificated
ESBWR	GE-Hitachi	1550	BWR	Certificated
US EPR	Areva NP	1600	PWR	Expected 2009
PBMR	Westinghouse	160	GCR	Expected 2012
IRIS	Westinghouse	350	PWR	Expected 2010
USAPWR	Mitsubishi	1700	PWR	Expected 2011
[1]ACR	AECL	1200	CANDU	Pre-certification

[1]Certificated in Canada

the Arab interest could be that the current high prices of oil and gas are permanent and offer greater foreign earnings and the replacement of old gas turbine combined heat and power plants with nuclear is one way of taking an indigenous fuel out of the domestic market to sell to the rest of the world.

They have all said that they will not enter the fuel cycle but that they will buy fuel through the reactor suppliers. Another oil producer with nuclear ambitions is Indonesia, which with Turkey has sought bids for a first reactor. Thailand is also likely to seek bids in the near future. Both Turkey and Egypt have been seeking nuclear power for more than thirty years. Turkey has nominated the site for a 4000 MW station at Mersin on the Mediterranean coast and a call for tenders was issued at the beginning of April 2008. Egypt meanwhile is reportedly studying tenders for a first reactor to be installed at El Dabba, west of Alexandria on the Mediterranean coast.

Would nuclear energy have been politically sabotaged in the United States and Europe if it had not first been used to make bombs which were dropped on two Japanese cities to end the Second Word War? This is something we will never know, but the fact is that over fifty years ago a system of electricity production was developed which had no emissions and the potential to supply electricity at low cost and high availability which is since proven.

Watts Bar is the last nuclear plant to be built under the old licensing system in the United States. The new licensing procedure combines construction and operation and is awarded for a certified reactor type and is granted for 20 years. Twelve licenses have already been issued.

5.5: Okiluoto, Finland: View of the site in August 2008 with the first EPR reactor, Unit 3, under construction at left. Application has been made for a fourth unit at the site. (Photo courtesy of TVO)

Of the new reactors in Table 5.1, the ABWR is already in operation in Japan, and the first four examples of the Westinghouse AP 1000 were ordered in the summer of 2007 from China with two units each on the Haiyang and Sanmen sites. For these they have commissioned Doosan Heavy Industries, of Korea, to supply pressure vessels, pressurisers and steam generators

EPR is the 1600 MW PWR developed in Europe. The first is under construction at Okiluoto, Finland, and scheduled to be in service at the end of 2012. Application for a fourth unit on the site was made to the Finnish Government in March 2008. Presumably the construction is planned to start before the Unit 3 work force has been fully dispersed from the site.

The second EPR is being built at Flamanville, France. Then in November 2007 a contract worth 8 billion Euros was awarded by the China Guangdong Nuclear Power Corporation for two EPR reactors and fuel for the first 15 years of operation. The reactors will be built on a new site at Taishan, in Guangdong province This is the largest contract that Areva had received and builds on the experience of previous contracts at Daya Bay and Ling Ao with a total of eight 1000 MW reactors on two sites.

Although there are only two reactors under construction in Europe, at least seven new units could be in operation by 2020. But for now, the main markets for nuclear power are China, Japan, and Korea, and also Russia, the United States, Canada and South Africa.

TABLE 5.2: CHINESE NUCLEAR PROGRAM TO 2015

Plant	Location	Type	Output MW	Supplier	Service Date
Quinshan I	Zhejiang	PWR	300	Mitsubishi	1991
Daya Bay	Guangdong	PWR	1000	Areva	1994
Daya Bay	Guangdong	PWR	1000	Areva	1994
Quinshan 2	Zhejiang	PWR`	600	CNPE	1997
Quinshan 3	Zhejiang	CANDU	728	AECL	2003
Quinshan 4	Zhejiang	CANDU	728	AECL	2003
Ling Ao 1	Guangdong	PWR	1000	Areva	2003
Ling Ao 2	Guangdong	PWR	1000	Areva	2004
Quinshan 5	Zhejiang	CPR1000	1080	CNPE	2004
Tianwan 1	Jiangsu	PWR	1000	Rosatom	2004
Tianwan 2	Jiangsu	PWR	1000	Rosatom	2005
Ling Ao 3	Guangdong	PWR	1000	Areva	2010
Ling Ao 4	Guangdong	PWR	1000	Areva	2011
Quinshan 6	Zhejiang	CNP 600	650	CNPE	2011
Hongyanhe1	Laoning	CPR1000	1080	CNPE	2012
Ningde 1	Fujian	CPR1000	1080	CNPE	2012
Quinshan 7	Zhejiang	CNP 600	650	CNPE	2013
Hongyanhe 2	Laoning	CPR1000	1080	CNPE	2013
Haiyang	Shandong	AP 1000	1100	Westinghouse	2013
Ningde 2	Fujian	CPR1000	1080	CNPE	2013
Sanmen	Zhejiang	AP 1000	1100	Westinghouse	2013
Haiyang	Shandong	AP 1000	1100	Westinghouse	2014
Hongyanhe 3	Laoning	CPR1000	1080	CNPE	2014
Ningde 3	Fujian	CPR1000	1080	CNPE	2014
Sanmen	Zheijang	AP 1000	1100	Westinghouse	2014
Taishan 1	Guangdong	EPR	1600	Areva	2014
Hongyanhe 4	Laoning	CPR1000	1080	CNPE	2015
Taishan 2	Guangdong	EPR	1600	Areva	2015
Ningde 4	Fujian	CPR1000	1080	CNPE	2015
Taishan 2	Guangdong	EPR	1600	Areva	2015

Of these China has the largest programme of nuclear power development, with eleven reactors in operation totaling 9436 MW and sixteen more under construction which will all be in operation by the end of 2015. Total nuclear capacity then will be 33 750 MW, by which time a further 18 000 MW will be under construction.

China is said by the Green Activists, who are painting a picture of them as a grossly polluted country, to be installing one new coal-fired power plant a week. The reality is that China plans to have 40,000 MW of nuclear plant in operation by 2020 which will account for 4% of the total generating capacity in the country, and a further 200 000 MW by 2050, many of which will be the new reactors certified in the United States and Europe.

The Chinese Government are acutely aware that old coal fired plants are responsible for 80% of the country's emissions and that but for pollution the Gross National Product (GDP) could be as much as 7% higher. Hence to meet growing demand for energy in the wake of industrial development, there is the big nuclear programme and also development of the Three Gorges (26 x700 MW) and other large hydro schemes, which have no emissions and will protect many more communities from flooding, and provide irrigation for improved crop yields, but which foreign Green factions vehemently oppose. In parallel with these clean energy developments, there is a broader plan of energy efficiency which aims to reduce the energy required to produce $100 worth of GDP by 4% per year.

China also aims to build it own nuclear plants, and is in fact the ninth country to develop its own indigenous design. The country's first reactor at Quinshan, a 300 MW PWR completed in 1994, was designed in cooperation with Mitsubishi who also provided some of the hardware. China Nuclear Power Engineering Company is the reactor design and production company which is gradually taking on more of the design and construction work. CPR 1000 is a Chinese designed three-loop PWR with a rating of 1080 MW. The first of four units are under construction at Hongyanhe in Laoning Province which will come into operation in 2012, with the others following at twelve month intervals.

China is also developing a gas-cooled reactor based on the German Pebble Bed Reactor technology. The prototype, designated HTR-10, is rated 10 MW, and went into operation at the beginning of 2007 at Tsinghua University, north of Beijing.

A 200 MW commercial prototype, of the Chinese gas-cooled reactor started construction a few weeks later with Huaneng Power International funding 50 % of the $300 million cost. The Chinese view is that with a planned nuclear capacity of more than 240 GW by 2050, they do not want water-cooled reactors alone, which depend on engineered safety sytems. The gas-cooled reactors are inherently safe and are of a standard design which can be factory assembled. As a 200 MW unit it could replace coal-fired capacity in many of the worst polluted cities.

Korea, with few indigenous fossil fuels has built nuclear plants for security of energy supply. They currently operate 20 reactors on five sites with a total capacity of 18 516 MW. Apart from four CANDU reactors at Wolsong, all the other reactors are PWR, from Westinghouse

TABLE 5.3: KOREAN NUCLEAR PLANS TO 2016

Reactor	Type	Supplier	Output MW	Service Date
Kori 1	PWR	Westinghouse	570	04/1978
Kori 2	PWR	Westinghouse	630	07/1983
Wolsong 1		AECL	635	04/1983
Kori 3	PWR	Westinghouse	950	09/1985
Kori 4	PWR	Westinghouse	950	04/1986
Yonggwang 1	System 80	Westinghouse	945	08/1986
Yonggwang 2	System 80	Westinghouse	945	06/1987
Ulchin 1	PWR	Areva	950	09/1988
Ulchin 2	PWR	Areva	950	09/1989
Yonggwang 3	System 80	Westinghouse	989	12/1995
Yonggwang 4	System 80	Westinghouse	989	03/1996
Wolsong 2	CANDU	AECL	680	07/1997
Wolsong 3	CANDU	AECL	680	07/1998
Wolsong 4	CANDU	AECL	680	10/1999
Ulchin 3	KSNP[1]	Doosan	995	08/1998
Ulchin 4	KSNP	Doosan	995	12/1999
Yonggwang 5	KSNP	Doosan	1000	05/2002
Yonggwang 6	KSNP	Doosan	1000	12/2002
Ulchin 5	KSNP	Doosan	1000	07/2004
Ulchin 6	KSNP	Doosan	1000	08/2005
Shin Kori 1	OPR-1000[2]	Doosan	1000	12/2010
Shin Kori 2	OPR-1000	Doosan	1000	12/2011
Shin Wolsong 1	OPR-1000	Doosan	1000	12/2011
Shin Wolsong 2	OPR-1000	Doosan	1000	10/2012
Shin Kori 3	APR-1400	Doosan	1350	09/2013
Shin Kori 4	APR-1400	Doosan	1350	09/2014
Shin Ulchin 1	APR-1400	Doosan	1350	12/2015
Shin Ulchin 2	APR-1400	Doosan	1350	12/2016

1 Korean built PWR based on Westinghouse System 80 design.
2 Optimised Power Reactor based on System 80+ for sale in Asian market.

and Areva. Average availability over all 20 reactors in 2006 was 92.3% compared with a world average of 79.5%.

Like China, Korea aims to design and build its own nuclear plants, but based on the former ABB Combustion Engineering System 80 design. With completion of their first plant, Kori 1, an 800 MW British supplied, Westinghouse PWR, in 1978, the engineers were all graduates of American Universities who had gone on to receive nuclear training and operating experience on plants in the United States.

Subsequently a growing number of Korean engineers and operators

5.6 Korea: Yonggwang; Four units at this station are the first Korean built System 80+ reactors, designated Type OPR1000. (Photo courtesy of KNPC)

have been trained at Korean Universities and on the Kori power plant. In a relatively short space of time there was created a large population of Korean nuclear engineers either trained in the United States or trained on Korean plants designed in the United States by people who had themselves been trained there.

Coupled with this was the building of a factory to manufacture the heavy steel components: pressure vessels, steam generators and pressurizers. That company became Doosan Heavy Industries who in addition to their nuclear work have licenses for the maufacture of heat recovery boilers for the combined cycle market. The first Korean manufactured reactors were Units 3 and 4 at Yonggwang based on the original Combustion Engineering System 80, a 2-loop PWR design rated at 1050 MW and known as Type OPR 1000. Two more units at Yonggwang came into operation in 2002.

This reactor will be offered to the Asian market, but initially eight are being built to this design as additions to existing sites, starting with four units at Shin Kori (shin being the Korean adjective corresponding to new in English) on which construction has already started on the first two for commercial operation in 2011 and 2012. These will be followed by two units at Shin Wolsong. Units 3 and 4 of Shin Kori and Units 1 and 2 of Shin Ulchin will be the enlarged design known as APR 1400, initially rated at 1350 MW. All these plants are designed as two units sharing certain common facilities.

Seven units will be in commercial operation by December 2015 with the eighth, Shin Ulchin 2 following within a year. At the end of 2016, the total nuclear capacity in Korea will then have reached approximately 27,000 MW from 28 reactors. For the future, Korea aims to achieve 60% nuclear electricity supply.

The lack of significant indigenous fuel resources has been the main driver for Korea which, with a continuing programme of nuclear construction has also invested in combined cycle with orders for large district heating plants at new town developments.

Japan, was the first Asian country to start nuclear power production with the purchase of a 160 MW Magnox gas-cooled reactor from the UK which went into service in 1966 and was shut down in 1998. Now the country has the largest operating fleet of the Asian countries, with 55 reactors supplying 30 percent of the national electricity supply Nine more are under construction or planned, which will raise the nuclear share to 40% by 2016.

Japan imports about 80% of its energy needs and forty years ago took the decision that nuclear energy was a matter of strategic national importance, and since they imported all their uranium it made sense to reprocess the spent fuel, and recover the unburned uranium and plutonium as mixed oxide fuel. That way they could get an extra 30% of energy from the original fuel charge.

From 1977 to 2006, Japan Atomic Energy Agency has operated a 90 t/year pilot plant at Tokai that has reprocessed some 1100 tons of spent fuel. Until now, however, the reprocessing of spent fuel has been done mainly in Europe. British Nuclear Fuels Ltd (BNFL) at Sellafield received 4200 t and Areva at Cap la Hague 2900 t.

Mixed oxide fuel assemblies are being produced in Europe and the vitrified high-level wastes will be returned to Japan for disposal. Areva's contract ended in 2005, while in Japan spent fuel has been stored at the power plants to await the start up of a new 800 t/year reprocessing plant at Rokkasho-mura in May 2008.

Rokkasho-mura in Aomori Prefecture in the north of Honshu is becoming the principal fuel cycle centre in Japan. Operated by Japan Nuclear Fuels Ltd, it has a centrifuge enrichment plant which began operating in 1992 with 300,000 swu/year and is working up to a final capacity of 1.5 million swu/year. The reprocessing plant will produce 4 t/year of fissile plutonium to make into mixed oxide fuel, and has a work load of 32 000 t of spent fuel from Japanese reactors to reprocess over the next 40 years.

A Mixed Oxide Fuel fabrication plant is now under construction. The nine Japanese Utilities who are the shareholders of JNFL have amassed a large inventory of reactor-grade plutonium to be made into mixed oxide fuel. more than 40 t is held in Japan with a further 14 t held in France and 11.3 t in the United Kingdom. This is no problem since it is a national policy to recycle plutonium as mixed oxide fuel in the existing power reactors.

Japan has also designed and built reactors from the original Westinghouse license, but only for the Japanese market. However Mitsubishi has supplied some equipment and assisted in the development of the Chinese CPR1000 reactor. They have also developed a 1700 MW PWR for which they seek US certification, and have two potential orders from Luminant Corporation, at Commanche Peak, Texas.

The other facility at Rokkasho-mura is the waste repositories for low level and high level nuclear waste. With the plutonium recycling policy, for every 100 t of spent fuel, 97 t is recoversd to be recycled and 3 t is high level waste. This material will be encapsulated in borosilicate glass blocks which will be consigned to the repository.

Two reactor developments arose from the fuel cycle work. First was Monju, a 250 MW sodium-cooled fast breeder ractor which went into operation in 1994. Two years later a sodium leak led to it being shut down, and it has not operated since. The project was taken over by JAEA and the Ministry of Science and Technology, and the repaired reactor was due to restart at the end of 2008.

A prototype gas cooled reactor, the 30 MWt High Temperature Engineering Test Reactor (HTTR) came into operation in 1998. This is a graphite-moderated and helium-cooled reactor. with an operating temperature of 950°C. It is designed for application to chemical processes such as thermochemical production of hydrogen. The fuel is ceramic-coated particles incorporated into hexagonal graphite prisms.

This reactor is designed to establish the commercial system for the second-generation helium-cooled plants running at high temperatures for either industrial applications or to drive direct cycle gas turbines. By 2015 an iodine-sulphur plant producing 1000 m^3/hr of hydrogen is expected to be linked to the HTTR to confirm the performance of an integrated production system.

Japanese progress as a major nucler power producer has come despite it being one of the world's most seismically active countries. Earthquakes have frequently led to the shut down of power plants, all of which have survived without damage and been brought back into service. Japanese earthquakes are generally not very powerful. But in 2006 a powerful earthquake some 200 km north of Tokyo had its epicentre a few kilometres from the Kashiwazaki Kariwa power station.

5.7: Kashiwazaki Kariwa, Japan: TEPCO's largest nuclear plant with two examples of GE's Advanced Boiling Water Reactor at far left. (Photo courtesy of TEPCO)

The staion immediately tripped out, and while inspection showed that the nuclear systems were not damaged, the buildings were, and a number of small fires have broken out over the last two years. In May 2009 the Prefectural Authorities gave consent for TEPCO to restart unit 7, one of the two GE Advanced Boiling Water Reactors on the site. This was due to start around May 15 and would be a series of tests at gradually increasing power levels. Once full power is reached there will be an official government inspection before the return to full commercial operation. If successful, TEPCO will see a ¥70 billion reduction in fuel purchases and a 5 million t/year reduction in greenhouse gas emissions.

Russia is a country with most of its population west of the Ural Mountains and most of its energy resources to the east in Siberia. In the time of the Soviet Union nuclear plants were installed in European Russia and Ukraine, and a number of units were sold to its client states in Eastern Europe, Cuba and also Finland.

The nuclear industry is currently being brought together into a single conglomerate Rosatom, for mining and enrichment of uranium, fabrication of fuel, building and maintaining nuclear power reactors, reprocessing spent fuel and disposal of nuclear waste. Also included in this is the nuclear plant operator Rosenergoatom.

The collapse of Communism was a traumatic time for the Russian nuclear industry. Put mildly they knew what they had to do but because they were not being paid for the electricity produced they had no money

5.8: Kalinin, Russia: the first station with the 1000 MW PWR went into service in 1984. There are three operating reactors on the site and a fourth under consturction. (Photo courtesy of Rosenergoatom)

to pay operators or perform maintenance or purchase fuel. Gradually things were brought back to normal with various government block payments over the years, in order to keep them running. But a ruling under the Yeltsin Administration called for any nuclear power station to be shut down if the condition of the plant became at all dangerous.

One of the first Acts under under the Putin Administration was to bring the nuclear operators into Rosenergoatom which was founded in 1992. A further reorganization in 2007 brought the entire nuclear power industry into the State Nuclear Power Company, Rosatom headquartered in Moscow. Various subsidiary companies deal with aspects of the fuel cycle, power plant construction, and operation, and reprocessing.

Currently there are in Russia ten operating sites with 31 reactors at a total output rating of 23 200 MW. Of the total nine are the VVER 1000 a PWR at nominally 1000 MW. These are at Balakovo (4) Kalinin (3); Novovoronezh (1), alongside four of the smaller VVER 440, installed between 1970 and 1981. Other sites with the VVER 440 reactor are Kola (4) and Volgadonsk (2). There are four 1000 MW RBMK reactors operating at Sosnovy Bor, near St Petersburg, and another four at Smolensk..

Russia operates the complete fuel cycle and supplies fuel and reprocessing services to export contracts. The 600 MW unit at Beloyarsk is now the only operating example of a liquid sodium-cooled fast breeder reactor, following the closure of the Creys Malville

5.9: Braidwood, IL, USA. Located 100 km southwest of Chicago, these two 1200 MW Westinghouse PWR's went into service in 1988. (Photo courtesy of Exlelon Corporation)

and Dounreay plants in France and the UK, respectively. It has been operating for some 28 years over which time it has generated more than 100 TWh of electricity. In 2007 it was shut down for the turbine and control systems to be upgraded and the original 30 year design life to 2010 has been extended to 2025.

Also in 2007 a joint venture was formed between Alstom and OAO Atomenergomasch to build the Alstom 1500 rev/min Arabelle turbine for Russian nuclear power plants. The first order was placed in September 2008 by Atomenergoproekt for a new plant to be built at Seversk, near Tomsk in Western Siberia.

This is the first of a series of new plants which will have the uprated version of the VVER 1000. This is rated at 1200 MW and and is designed as a standard nuclear power plant known as AES 2006. The VVER 1000 is also the reactor model which is being offered for export. Two have been supplied to the Tianwan project north of Shanghai in the Jiangsu Province of China, and two more to Temelin in the Czech Republic. Discussions are in progress with the Government of Bangladesh for the construction of two 1000 MW reactors in the country,

The United States is treading very cautiously to keep the nuclear option open and during the last eight years of the George W Bush presidency there has been a concerted effort to revive the nuclear industry after the collapse engineered by his predecessor Jimmy Carter, in 1979. Not only have new reactor designs been certificated,

5.10: Susquehanna, PA, USA: PPL's twin BWR station has had turbines uprated and license extension. Plan to install a US-EPR nearby at Bell Bend. (Photo courtesy PPL)

but licenses have been issued for new construction for plants to be in operation by 2017 at the earliest.

The ban on reprocessing and the operation of the once through fuel cycle enforced at that time continues. But under the Bush Administration the nuclear licensing procedures have been revised and New Start licenses have been granted to several utilities. while there have been many plants that have had their operating licenses extended to 60 years. But in the background are the developments of new reactor designs which will carry the benefits of nuclear power to smaller countries and Island networks which cannot support a 1000 MW or greater power plant on their system.

Three designs have evolved for smaller reactors which can fit into smaller networks and also be installed as several parallel units to meet load growth in a particular area. The common features of all three reactors are a low output, higher fuel enrichment to achieve longer intervals between refueling, and a 60 year design life.

First of these is the Pebble-Bed Modular Reactor (PBMR) a high-temperature gas-cooled reactor which was originally developed in Germany and was taken up in South Africa. This is rated at 160 MW with a direct closed-cycle gas turbine drive. It uses uranium dioxide fuel enriched to 10% which stays in the reactor for up to six years.

The second reactor, IRIS, is an international project to develop a smaller, simplified pressurized water reactor with the heat exchangers and pressurizer contained within a single pressure vessel. The initial design is rated at 330 MW, and it also has higher fuel enrichment which can stay in the reactor up to four years.

The third reactor is also a pressurized water reactor, which is a Babcock & Wilcox design for a modular, factory-assembled unit rated at 125 MW. The pressure vessel contains a single helical steam generator above the core and the pressurizer is incorporated in the vessel head. The reactor uses the standard PWR fuel at 5% enrichment which will stay in the reactor for four years between refuelling outages

The particular feature is that the reactor will be sited under ground in a containment module and several could be mounted side-by-side with just the steam connections to and the condensate returns from the turbine, which with the control room will be the only part of the station above ground. Several reactors could be installed on one site to meet load growth as it happens.

PBMR and IRIS have license applications before the Nuclear Regulatory Commission. Babcock plan to file an application for a four-unit demonstration plant in 2011 with a likely in-service date of 2018. Tennessee Valley Authority have been in discussion with Babcock over a possible site.

At present the country has the largest nuclear power fleet in the world with 104 operating reactors and one under construction. Nuclear plants produce 20% of the country's electricity supply.

Among the twelve New Start licenses issued, are those to (TVA) at their Bellefonte site for the first Westinghouse APR1000, and Entergy's Grand Gulf, Mississippi, site which will have the first of the GE ESBWR.

PPL Corporation have contracted with UniStar Nuclear, a joint venture of Constellation Energy and Electricité de France, who have applied for certification of Areva's EPR design which if successful will be the first foreign-designed nuclear reactor to be installed in the United States. The aim is to apply for a New Start license for Big Bend, nearby their existing Susquehanna site.

There PPL have two 1100 MW BWR's which entered service in 1983 and 1985. Since then the operating license has been extended to 60 years and the steam turbines have had low pressure section upgrades giving each an extra 50 MW of output.

The only new addition to the nuclear fleet in progress is TVA's second unit at Watt's Bar, a 1200 MW PWR which is planned to be in operation in 2013. But in April 2008 Georgia Power signed a contract with Westinghouse and The Shaw Group for the engineering, procurement and construction of two AP1000 reactors and related facilities.

Georgia Power is one of four shareholders in the existing Vogtle nuclear station south of Augusta, GA, which currently has two 1200 MW Westinghouse PWR's which came into operation in June 1987 and May 1989. The new units will be built beside the existing station as units 3 and 4. Planning consents were concluded at the end of March 2009 so that work can start for both units to be in service by the end of 2017.

Since then Westinghouse, in partnership with Shaw Group, has received orders for another eight units at four sites, each with two APR 1000 reactors. This will see the two Chinese plants in operation by 2014, with three US sites in South Carolina, Georgia, and Florida in operation by 2018.

Nuclear power development slowed down in the 1980's and the industry was left to depend on exports and service contracts on existing power plants. Even then, the Carter ban on reprocessing left foreign operators fearful that a new contract with a US supplier might include clauses to restrict operation to a once-through fuel cycle, and other than in Japan and Korea, the Franco-German company Areva, and Atomic Energy of Canada, have supplied most of the few nuclear plants built in the last twenty years.

Atomic Energy of Canada have redesigned their CANDU reactor as ACR1000 a nominally 1200 MW unit. The basic design follows that of the existing CANDU reactors except that light water is used as the primary coolant and extended fuel life is obtained with a low enrichment. The heavy water is only used as a moderator and the reactor contains about 60% less by volume. The reactor has been designed to use mixed oxide and thorium fuels and like all previous CANDU designs it can be refueled on load.

The Ontario Provincial Government's policy of July 2005 had announced the intention to close the four remaining coal-fired plants in the province on grounds of public health and replace them with two nuclear stations and combined cycles for mid-load duty and frequency control. Following this announcement the Canadian nuclear market has come alive again with inquiries from the provinces of Alberta and New Brunswick as well as the two plants planned for Ontario.

Ontario Power Generation with Pickering (8 x 540 MW) and Darlington (4 x 935 MW) is the largest nuclear operator in North America and has offered Darlington for one of the sites. The other nuclear operator in Ontario is Bruce Power owned by a consortium of Cameco Corporation, Trans Canada and BPC Public Generation Trust. This company owns the two Bruce Power Stations which together provide 20% of the Ontario electricity supply. The company have room on their site for a third station of four units.

New Brunswick is seeking another nuclear plant to enable them to export low cost power to the New England states. In Alberta, Bruce Power has picked a site at White Mud, southwest of Edmonton for a 2-unit station which could ultimately be extended to four CANDU ACR 1000 reactors.

In January 2008 the British Government delivered a new energy policy which provides for the construction of initially up to ten new reactors from currently available designs. Electricité de France (EdF) plans to build four of the 1600 MW EPR at existing sites. Other designs to be considered were the Westinghouse AP 1000 and the GE-Hitachi ESBWR and the CANDU ACR 1000; but in 2008 AECL pulled out of the bidding to handle a growing domestic market

What is interesting is that no limit has been placed on the role of nuclear power, which suggests that they will not only replace the reactors now operating but some or all of the coal-fired power plant as well. In fact the White Paper gave the clearest indication yet that the aim was to bring to an end coal-fired power generation in order to achieve the drastic cut in greenhouse gas emissions (80%) proposed in the targets for 2050.

At that time the British Government held 35% of the nuclear operator British Energy, which has fifteen reactors on eight sites and a 2000 MW coal fired power station at Eggborough. Bids were invited for the company and EdF had its bid accepted at the third attempt in September 2008. EdF are the world's largest nuclear operator with a large customer base in London and southeast England managed by a British registered company, EdF Energy.

EdF's initial bid was rejected by the British Energy Board but a new bid worth £12 billion was accepted. The British Government received £4.2 billion as a cash payment for their shareholding. Centrica will hold 20% of British Energy and will receive 20% of the output of each of the first four reactors.

Much of the irrational criticism of the deal has gone now that it is done and with EdF's supplier, Areva, being the only Western European nuclear constructor. Also by standardizing on the EPR, the problem which has bedevilled past nuclear programmes is no longer there. There are a number of British companies that have supplied engineering services to Areva's contracts and will no doubt provide a significant British content to the projects.

The British Government has said that it will create favourable

TABLE 5.4: BRITISH NUCLEAR SITES

Site	Current status `	Output MW	Developer
Berkeley	Decommissioned Magnox		NDA
Bradwell	Decommissioned Magnox		EdF Development
Braystones	New site for PWR		RWE Npower
Chapelcross	Decommissioned Magnox		NDA
Dungeness	Operational AGR	1320	British Energy
Hartlepool	Operational AGR	1320	British Energy
Heysham	Operational AGR	1320	British Energy
Hinkley point	Operational AGR	1320	British Energy
Hunterston	Operational AGR	1320	British Energy
Kirksanton	New site for PWR		RWE Npower
Oldbury	Operational Magnox	600	E.ON/RWE Npower
Sellafield	Decommissioned Magnox		NDA
Sizewell	Operational PWR	1100	British Energy
Torness	Operational AGR	1320	British Energy
Trawsfynydd	Decommissioned Magnox		NDA
Wylfa	Operational Magnox	600	E.ON/RWE Npower

conditions for licensing new nuclear plants and has already introduced a bill to speed up applications for consent to build large infrastructure facilities in the public interest, which have in recent years become bogged down in protests.

In April 2009 the Government published a list of eleven sites for new nuclear power stations. Eight of these are the existing sites in England and Wales with another three in Cumbria at Sellafield, Kirksanton and Braystones. All sites are expected to have pressurized water reactors, the first four of which are to be built at Hinkley Point and Sizewell. These will be two 1600 MW EPR's, on each site, with the first to be completed at Hinkley Point at the end of 2017.

If all eleven sites are developed then 18400 MW of new nuclear and at least 11000 MW of new combined cycle capacity would be in service by 2023. with Sizewell B as the only one of the current nuclear plants still operating and Drax the only remaining coal-fired plant

When British Energy was privatized the original 26 Magnox reactors, many of which were on the point of shutting down were made over to British Nuclear Fuels, out of which was created the National Decommissioning Authority (NDA).

NDA has all the sites with only Magnox units on them, and British Energy those sites with operating reactors on them. The wholly NDA sites are Berkeley, Bradwell, Chapelcross, Oldbury, Sellafield, and Wylfa. Additional land at three sites has been sold to EdF Development, 100 hA at Bradwell, and to a joint-venture of E.ON (UK) and RWE Npower, for 48 hA at Oldbury and 178 hA at Wylfa, which will be the last of the Magnox stations to shut down at the end of 2010.

In May 2009 Centrica completed a deal with EdF for £1.1 billion in cash and a 20% share in British Energy which also includes entitlement to 20% of the output of the first four of the new EPR reactors. EdF will supply 18 TWh to Centrica over five years from 2011. In return EdF acquired Centrica's 51% share in the Belgian Generator SPE.

Under this arrangement Centrica with at present only gas-fired capacity supplying 58% of its energy gets access to additional nuclear supplies which help to protect it against future fluctuations in wholesale gas prices.

Of the other five companies that dominate the Electricity market in the UK, only E.ON has decided on a specific reactor type. In April 2008 they entered an agreement with Areva under which they will take the 1600 MW EPR reactor as a standard unit for their British and other sites around the world.

EPR is a Franco-German design based primarily on the latest PWR design in service in France. This is the 1400 MW Type N4, of which four are in service with two each at Chooz and Civaux.

Currently there are 59 operating reactors in France with one under construction at Flamanville on the Cherbourg Peninsular. This is the second EPR unit which is being built alongside two 1300 MW units dating from 1986. EdF recently have ordered a second EPR to be built at Penly on the Channel coast some 50 km north of Dieppe.

The first of the 900 MW PWR was installed at Fessenheim, in Alsace, in 1977. The last of 32 reactors of this type were Gravelines 5 and 6 which were timed to completion of the cross-channel HVDC circuits in 1985 and 1986 so that their output could be fed to London and the south of England. EdF later took over the two regional distribution companies; London Electricity and Southeastern Electricity

France will soon have to start decommissioning some of their early PWR's and in time these will be replaced with up to forty of the EPR. But the programme will not be defined until the unit at Flamanville which is due to enter service at the end of 2013 has been in operation for a few years.

South Africa is particularly interesting as it took over development of a gas-cooled reactor design from Germany in 1990 after Green protest there had effectively expelled it. South Africa is a country with energy

5.11: Pelindaba, South Africa, near Johannesburg, is the development centre for the Pebble-Bed gas-cooled reactor. (Photo courtesy of PBMR Ltd)

resources in the wrong place, no indigenous gas fields, and limited hydro potential on the Orange and Vaal rivers. It imports hydro power via an HVDC line from the Caborra Bassa dam on the Zambesi River in Mozambique.

All the coal fields are in the north of the country around Johannesburg and there are some large coal-fired power stations in the same area with long transmission lines feeding south. The one exception is the Koeburg Nuclear Station situated on the Atlantic coast about 100 km north of Capetown. This has two 985 MW PWR, supplied by Areva, which have been in commercial operation since 1984 and 1985, respectively.

South Africa's problem is that to reduce pollution from power generation in the face of rising economic growth and energy demand, they have no natural gas. The majority of the population lives on the coast in Natal and Cape Province and around Johannesburg. Koeberg when it was completed in1986 put a large block of generating capacity in the South, where it was needed.

So the answer is nuclear power but in smaller units based on German technology of the Pebble-Bed Modular Reactor (PBMR) This is a gas-cooled reactor which is rated at 160 MW with a closed cycle gas turbine as the output device.

The origin of PBMR was a 15 MW gas-cooled reactor built at the Jülich Nuclear Research Centre about 20 km north of Aachen near the Belgian frontier. It was conceived as a small intrinsically safe reactor of a size which could be installed anywhere in the world, so that everybody could benefit from a clean emission-free power source. At the time that the reactor came into operation in 1973 there was mounting concern about acid rain which would be completely avoided with an emission-free nuclear power plant. It was clear, too that the German developers saw it as a reactor which could be exported all round the world

This was a high temperature reactor with an output gas temperature of 950°C. When it was being built the first of the 100 MW class gas turbines running at synchronous speed were under development which also had turbine inlet temperatures of about 900°C, and there were coal-fired closed-cycle gas turbines, operating at some German district heating plants.

The reactor powered a steam turbine but the long term aim was to carry commercial development to a closed cycle gas turbine output. In fact a coal-fired closed cycle gas turbine using helium was installed at the Oberhausen district heating plant in 1976.

The German plan was to use thorium carbide as the fuel and a scaled up version of the reactor at Jülich was built at Hamm-Uentrop but only operated for two years before Green entry into German state politics put an end to nuclear construction and new reactor development. By 1990 the Pebble Bed Technology was at an end, but there were other people who had studied the reactor concept and were convinced of its advantages, and that it had the potential for a wider range of industrial applications.

The advantage of PBMR is that it is gas-cooled and therefore intrinsically safe. It is a small reactor which is housed in a steel reactor vessel which is only 27 m high by 6.2 m diameter. It is therefore of a size which can be factory assembled

South Africa is also a uranium producer and the view of the Government is that they should exploit an indigenous resource which is the only one that can help them to cut emissions. The coal-fired plants all date from the 1980's and earlier and all operate on the subcritial steam cycle. Several of them have been refurbished and brought back into operation to meet increased electricity demand on the back of rapid industrial growth.

The developers of PBMR have secured land adjacent to the Koeburg power station for a prototype unit which could be completed by 2014. But after that Eskom, the State power utility, has placed an order for initially 14 units, most of which are for installation at different locations around the country.

A big difference between the PBMR and a conventional PWR apart from the size is the fact that it has a bigger load following capability and is claimed to be able to operate down to 20% of full load. Once proven in service, this gives a new dimension to nuclear power which could also supply mid-load power, for example, running down to 50% overnight and shutting down at weekends

Westinghouse Nuclear, now a Division of Toshiba Corporation, of Japan were formerly owned by British Nuclear Fuels, who are a shareholder in the PBMR Company. As the reactor engineering company they took over the shareholding through their time at BNFL, and now are working for certification of the reactor in the United States. A prototype is to be built at Hanford, WA.

But in South Africa, as the performance of the reactor is disseminated around the world, some of the later units are reserved for export sales. The high exit temperature of the reactor at 900°C would make it useful as an energy source for a coal gasifier. In a version with a steam turbine output PBMR could be an energy source for a large industrial combined heat and power or district heating scheme.

Why then should we consider nuclear to be green energy? It has no greenhouse gas emissions and relatively low environmental impact. The high fuel density means that the supporting energy loads of the power station are much smaller, and the reaction produces plutonium which can be made into mixed-oxide fuels for the same reactor.

It is the United States that has licensed the next generation of reactors, primarily it would seem to supply the customers in the Far East. It is orders from Korea, Japan, and China that have kept the American nuclear industry alive. The APR 1000 has been certificated; eight await construction and operation licences in the United States, and four have been sold to China.

It is worth noting that no country supplied with a reactor by Russia has ever developed and detonated their own nuclear weapon. By supplying all the fuel and calling back all spent fuel, Russia has kept a tight control on the fuel cycle and is in fact the only country currently operating a fast breeder reactor on a plutonium fuel cycle.

Lithuania has been pressured by the European Union into closing the two RBMK reactors at Ignalina of which one was shut down in 2003 and the other is due to close at the end of 2009. Lithuania is however in dispute with the European Union since they want to keep Ignalina running until a new reactor can be completed in about 2015. This is an international project, with Lithuania holding 34%, and Estonia, Latvia, and Poland with 22% each. The new plant will have two reactors, probably the 1600 MW EPR.

In Asia, besides India and Bangladesh, Indonesia, Thailand and Vietnam are looking at nuclear power. Indonesia is in discussion with Korea for a first nuclear plant in central Java. In Thailand, EGAT is reportedly looking for up to four reactors. which would all be water cooled reactors in the new designs. Vietnam, similarly' is looking for initially four 1000 MW reactors at two sites with the first in operation in 2021.

Gas-cooled reactors under development in South Africa, China, and the United States bring a new dimension to nuclear power, literally, with the South African design being for 160 MW output, and a Chinese commercial prototype at 200 MW on which work stated in 2008. Apart from the Japanese unit, the only other gas-cooled reactors in operation are the fourteen Advanced Gas-Cooled Reactors in the UK and the last two Magnox units.

A lot hangs on PBMR, development of which has been delayed by some forty years by Green anti-nuclear campaigning in Germany. If it can be proven to be flexible in operation then it is not only able to perform the same load following duties as a combined cycle, but could also be installed in the smaller networks, for instance on large islands which could support the 160 MW unit but not a 1000 MW plant.

There has been much concern over safety issues in the past, which have largely been solved. The only working Fast Breeder Reactor is in eastern Siberia at Beloyarsk but is currently shut down for maintenance and upgrading to extend its life for another fifteen years.

An extension of nuclear energy will require additional reprocessing facilities. At present the only commercial facilities are in the UK at Sellafield, Cap la Haag, France, Chelyabinsk, Russia and Rokassho-Mura in Japan. Of the existing nuclear operators, only a few are committed to reprocessing and recovering plutonium to make mixed-oxide fuel

Creation of the Global Nuclear Energy Partnership offers the chance to put the fuel cycle under control of a single global authority embracing all the nuclear designers and the fuel services, which should make for a broader acceptance of nuclear energy around the world.

The fact is that unless there is a dramatic reduction in the demand for electricity around the world, which is highly unlikely, nuclear energy is the only thermal system that can provide the electricity in the quantity that we need and emit no greenhouse gases while doing it.

6
Combined cycle

The gas-fired combined cycle has, since 1990, been the preferred power plant for system expansion over much of the world. It has low environmental impact, high efficiency and is quick to build. Its inherent flexibility of operation means that it can serve as a base- or mid-load or peaking unit at different times in its service life, as many of them are.

But for the future the combined cycle must be designed to play a specific role in the green energy system of tomorrow. It could be the only remaining fossil-fired option and has the highest efficiency of any thermal generating system. Considering the other power plants that will be used: nuclear is primarily a base load system, though future gas-cooled reactor designs will be more flexible; hydro started as base load, but flexibility of operation means that it is now mainly used for mid and peak loads; wind has low availability and can only operate within a given range of wind speed.

So the combined cycle will be required to fill gaps in supply from the other systems. It will cover for refueling and maintenance outages on the nuclear plants, it will also operate more in years after periods of low rainfall when hydro output is low; and it will cover periods when the wind is not blowing. It will continue to be the power plant for industrial combined heat and power, and district heating schemes. for some years to come.

The basic combined cycle comprises one or more gas turbines, each exhausting into a heat recovery boiler generating steam which is fed to a single steam turbine. As long as the gas turbine exhaust temperature is more than 500°C, no more fuel need be burnt to create suitable steam conditions, and the ratio of gas turbine to steam turbine output is approximately 2:1, so that a gas turbine of 280 MW would power a 130 MW steam turbine.

Progressive development over the years of the heavy-frame gas turbines for land-based use has resulted in the steady increase of efficiency of the combined cycle from 42% of the earliest multi-shaft examples installed around 1970, to 58% with the current single-shaft blocks, and the promise of 60% from the new generation of gas turbines currently under development.

Nearly thirty years of commercial operation around the world has shown that the combined cycle is the ideal system to achieve an immediate reduction in emissions from power generation. If, for example, a 40-year old 1200 MW coal-fired power station is demolished and replaced with combined cycles of the same total capacity, gaseous emissions will be about 40% less in volume, and since the fuel is methane 44% of the emissions by weight will be water vapour. The efficiency will be about 58% as compared with up to 36% at best for a coal-fired plant of the 1960's.

The combined cycle with three 400 MW single-shaft blocks would be much smaller on the same site. Each block would occupy an area of about 2hA, and if built in the United Kingdom, the top of each stack would be 60 m above the ground and not 200 m as for the large concrete multi-flue stack of the coal-fired station.

The cooling load will be significantly less since the total steam turbine capacity will be only one third that of the old steam plant. If the original station had been cooled by six large, hyperbolic, natural-draught cooling towers, these would have been demolished with the old station and replaced with a wet mechanical-draught system or an air condenser for each block of the new plant.

Fifteen years after the end of the Second World War, following the discovery of natural gas at Groningen, in the northeastern Netherlands, the first combined cycles were starting to appear in Europe. They were one of several developments aimed at improving the efficiency of power generation, which was then typically less than 30%, causing some concern as electricity demand increased rapidly with post-war reconstruction and rising living standards.

Various concepts were studied of linking gas turbines to the feed water heaters of large steam plants which raised efficiency to over 40%, but these used relatively small gas turbines with exhaust temperatures of less than 400°C. Several schemes of this type were built in the United States but, although they were the most efficient power plants in the country at the time, they did not catch on with the utility industry and only six were built.

But in Europe the 75 MW power station completed in 1960 at Korneuburg, Austria, was of a different design with two 25 MW gas turbines each with its own heat recovery boiler feeding into a common range supplying a steam turbine. Because of the low exhaust temperature of the gas turbines, burners had to be installed in their exhaust ducts to raise the steam temperature leaving the boilers. Since the gas turbines did not run at the synchronous speed they had to drive their generators through a reduction gearbox.

This first combined cycle plant had an efficiency of 32.5% which was a definite improvement, compared with that of a 75 MW steam turbine of the time operating on a low pressure, non-reheat steam cycle, which would have been about 28%. Korneuburg A went into service in 1960 and in base load operation, averaged 6000 h/year for the next 14 years.

But for this technology to improve there had to be more powerful gas turbines capable of running at synchronous speed. Higher exhaust temperatures would remove the need for supplementary firing, and if the gas turbine ran at synchronous speed it would not need a gearbox between it and the generator.

The first gas turbines to run at the 60 Hz synchronous speed of 3600 rev/min had been launched in the United States in 1968. Five years later, at the time of the 1973 oil crisis there were available at least five gas turbine designs of 75 to 100 MW capacity, and running at the common synchronous speeds, of which three were for 50 Hz networks turning at 3000 rev/min.

Some of the early combined cycles were created by repowering steam plants, a number of which in the years immediately after the war had been built with small turbines rated at between 30 and 50 MW. Some of these had been converted from coal to oil firing. Repowering entailed replacing the fired boiler with a gas turbine and heat recovery boiler.

Since the steam conditions from the new boiler had to meet the requirements of the steam turbine it was inevitably a compromise, and although a higher efficiency resulted it was not as high as it would be if the steam turbine had been designed to match the heat output of the gas turbine.

In 1976 the first combined cycle in Europe with one of the large 3000 rev/min gas turbines went into operation at the Drogenbos power plant in the southern suburbs of Brussels. A 90 MW Westinghouse gas turbine, which was already on the site, was fitted with a vertical heat recovery boiler from Cockerill Mechanical Industries (CMI) to replace the fired boiler serving one of the plant's three 37 MW steam turbines. From the beginning it ran in support of a growing nuclear base load, with daily start and stop and shut down at weekends. It maintained this pattern of operation for the first ten years, and was still available as a stand by unit on the Belgian system in 2000.

The development of the combined cycle has followed the development of the gas turbine. Larger gas turbines had higher exhaust temperature and mass flow and could therefore generate more steam at higher pressure and temperature. The steam turbine, moreover, could be designed to maximize the heat output from the boiler with a second low-pressure steam input.

As the gas turbine outputs increased it would be possible by 1990 to have two 120 MW gas turbines supporting a steam turbine of about the same capacity, with a two-pressure steam cycle and an efficiency of about 49%. But although the gas turbines had been developed in Europe and North America, the combined cycle market had not initially developed there but in Southeast Asia and the Asian Pacific countries.

In the United States combined cycle application took a different turn with the Public Utilities Regulatory Policy Act, PURPA, of 1979. For the first time this opened the market to independent combined heat and power schemes by guaranteeing the current market price for any excess power sold to the public utility. To qualify the plant had to have a steam host, which could be a large paper mill or chemical plant, anything, in fact, which had a large process heat or air-conditioning load, such as an airport terminal. It also provided a solid industrial gas turbine market in the United States from which other developments have followed.

In Europe much of the coal gas production had hitherto been used in the domestic market for cooking and space heating, which was a sensitive political issue on price and availability. The decision was taken to convert all existing coal gas installations to natural gas and, with availability of the resource as then known, the use of natural gas for power generation was limited.

During the 1970's people became aware of acid rain, which was attributed to the sulphur oxides in the flue gases released in the combustion of many coals. Flue gas desulphurization (FGD) systems were developed and were installed initially on new coal-fired power plants in the United States and Germany. The first exhaust catalysts were being fitted to cars to reduce nitrogen oxides, which also required unleaded fuels. The widespread use of gas-fired central heating in new housing saw the gradual removal of coal from the space heating market. So at this time there could be seen the beginning of an environmental clean up.

As part of this trend during the 1980's the gas turbine industry was concentrating on the development of low emission combustors. The aim was to remove nitrogen oxides formed from the oxidization of the excess combustion air at firing temperatures above about 850°C. Much lower emissions of nitrogen oxides (NOx) down to less than 10 vppm were reported. But more significantly, by the end of the decade, a contract principle was established whereby all gas turbine plants would guarantee NOx emissions on dry gas of 25 vppm, and on oil at 42 vppm with water or steam injection.

The basic environmental measures had been established for the combined cycle with the gas turbines of the time, and with the reduced competitiveness of coal due mainly to the compulsory fitting of FGD to all new power plants, the combined cycle at this time became the preferred equipment for capacity addition. The contract condition for NOx emissions continues to the present day, and there is only one gas turbine, Siemens' 45 MW model SGT800, which can meet the NOx condition for oil without water or steam injection.

All through the 1970's and 80's new gas fields were being discovered and brought into production in the North Sea, Indian Ocean, South China Sea, in and around Australia and New Zealand, South America, North Africa and Siberia. Natural gas comes into the 48 lower states from Alaska and central Canada.

Many of these new resources were near to economically developing regions of the world. Europe and North America had seen their rate of growth in electricity demand fall in the decade after 1973, but in the developing economies of southeast Asia, rates were growing at over 10% and rising.

As the leading economic power in the region, Japan, together with Korea and Taiwan, had to import all their fuel supplies. The use of gas for power generation with a system of high efficiency which could be built quickly was particularly attractive. Already by 1980 all three countries were importing liquefied natural gas (LNG) from Australia, Indonesia and Abu Dhabi. In Japan, there was a gas turbine industry which had been built on American and British licenses to the four major power engineering companies..

The first combined cycle in Japan was at Tohoku Electric Power Company's Higashi Niigata site on the Sea of Japan coast about 300 km north of Tokyo. It consisted of two 545 MW blocks each with three Mitsubishi Type M701D gas turbines and a steam turbine, which went into operation at the end of 1983. The six gas turbines of this project, which were then the largest in the world at 133 MW, were the first to have dry, low-emissions combustors.

6.1: Angleur, Belgium. Sulzer N1 10 gas turbine repowered a 30 MW steam turbine in 1967 and is still available as a stand-by unit on the Belgian system. (Photo courtesy of CMI)

In Taiwan the first combined cycle was at Tung Hsio, on the west coast of the island, about 150 km south of the capital Taipei, with three 400 MW blocks initially burning heavy oil. These would in later years be converted to natural gas, and two more blocks would be added

At around the same time in Malaysia, Tenaga Nasional Bhd were building a 600 MW combined cycle on the east coast of the peninsula at Paka. This had two blocks each with two GE Frame 9E gas turbines and one steam turbine. While in Thailand, the Electricity Generating Authority was completing their first combined cycle at Bang Pakong with two 450 MW blocks, each with four Siemens Model V94 gas turbines and a steam turbine.

By 1990 when deregulation of electricity supply started in the United Kingdom and opened up the European combined cycle market it was fair to say that the greatest experience of the new technology was to be found in Asia. The majority of combined cycles operating in the United States had been installed under the PURPA regulations as combined heat and power schemes for large process industries.

When the British Government started to privatize the electricity supply system the coal gas market had been completely converted for more than ten years. The amount of gas available to Europe had increased substantially, with piped supplies from the British and Norwegian sectors of the North Sea, and from Siberia coming in

6.2: Roosecote, UK: 220 MW combined cycle plant was built in the turbine hall of a former 120 MW power station near Barrow in Furness. (Photo courtesy of Centrica)

through Germany and Austria, and LNG from Algeria initially entering Spain at Barcelona..

The British privatization had two effects, first that combined cycles were installed which led to the closure of a lot of the older and smaller coal-fired plants, with consequently a significant reduction in greenhouse gas emissions.

Second, was the unlocking of the market for industrial combined heat and power. If anybody could generate power and sell it into the grid, then an industry could build its own power plant to provide some or all of its electricity and all of its process steam and, since it would need a back-up power supply to cover maintenance, there would be a permanent connection to the grid which would also enable them to export surplus power.

But this was not an exclusively British phenomenon. As deregulation spread around the world, in more countries the combined heat and power market sprang into life. Before this it was almost always the case that monolithic State-owned electric utilities had resisted demands for combined heat and power as a way to improve energy efficiency. Their objection was that if they built such a plant to sell steam it would mean that they would produce less electricity and could not earn as much from the sale of steam as from the sale of electricity.

What had changed was that generation had been separated from transmission and distribution, so that the companies running the grid system could buy power from anybody according to the output, price and availability of their power plant.

For example, in one case, there could be a 180 MW combined cycle with two 60 MW gas turbines and a steam turbine, selling all its energy through the grid to the distribution companies; and in the other case a similar plant owned by, say a large pulp and paper mill, which could supply all its process steam from the steam turbine without the need for supplementary firing and would have more than 140 MW of power, excess to requirements to sell into the grid for more than 6000 hours a year. Furthermore the efficiency of the combined heat and power plant would probably be at least 70% compared with 56% for the combined cycle.

While the British privatization was followed in several other countries, some of them separated transmission and distribution from generation and invited foreign investors in to build new generating plant, whilst retaining control of the grid. The 1990's saw a steady market for combined cycles around the world. Gas was cheap, and many of these plants were built for base-load duty. Economic recession in the Asian Pacific region slowed demand there after 1997 but since then growth has continued albeit at a slower rate. It is in China and India today that the double digit rates of growth in energy demand are occurring.

The United States were one of the last to deregulate and prospective generating companies bought into British and other power companies to gain experience of the liberalized power market. Similarly European generating companies and others went into North America.

For a time, between 1999 and 2002 it seemed that there would be a surge in the United States market which would see a large number of combined cycles installed to replace a large inventory of old coal-fired plant, much of which was more than 40 years old.

But at the end of 2001, American plans came to a shuddering halt with the bankruptcy of Enron, one of the leading gas and electric energy companies, who also had made a number of investments in other countries, including one of the world's largest combined heat and power plants in Northeast England, rated at 1800 MW and supplying process steam and electricity to the ICI Teeside chemical works.

The gas-fired combined cycle has firstly improved efficiency of energy supply, even with the combined cycle alone, and achieved what many people have been campaigning for in combined heat and power since 1973 and before. Every coal- or oil-fired steam plant that is replaced by a combined cycle or an industrial combined heat and power scheme results in an immediate reduction in greenhouse gas emissions from power generation. But for how long can it go on?

There is still plenty of gas in the world, and still gas is being flared off from some oil fields, although much of this has been stopped in the North Sea and elsewhere, where the gas can be easily recovered and sold. With the passage of time combined cycles have become more efficient, and there is the prospect of 60% thermal efficiency for electricity generation alone being achieved before 2010.

The development of the combined cycle has occurred in four separate stages. First were the so-called E-class gas turbines, which were launched between 1968 and 1980. Their particular property was that they all ran at the common synchronous speeds. By 1990 they had all been progressively upgraded to their final ratings between 95 and 150 MW, in 50 Hz ratings, and for all of them dry low emissions combustors were available.

The F-class gas turbines announced in the United States in 1989 and in Europe in 1992 were essentially larger machines with in some cases major design changes to reduce the cooling load of the gas turbine fabric so as to have more of the air for flame dilution to retain the low emissions compatible with those achieved in the earlier machines.

For the two European manufacturers the outboard combustor silos of the E-class machines were replaced with a full annular combustor similar to that found in the large turbo-fan aero engines. Dry low-emission combustors were now standard fittings to all the gas turbines used for power generation. Higher firing temperatures and mass flow, with exhaust temperatures of around 600°C, enabled them to support a tri-pressure reheat steam cycle with high pressure steam conditions close to those of the large sub-critical steam plants of the 1960's.

Both European manufacturers simultaneously launched models for 50 and 60Hz systems scaled from a common design base. Taking this a step further, smaller F-class machines were developed by GE and Siemens at about 60 MW with the scaled blade technology of the big machines and the same combustors and firing temperatures. These could not run at synchronous speed and therefore could drive through a gearbox for both 50 and 60 Hz systems.

It was also with the arrival of the F-class gas turbines that the move to a single-shaft configuration began with the gas turbine and steam turbine driving a common generator. The single-shaft block is now the standard combined cycle format in the 50 Hz market. It is typically erected and commissioned in between 24 and 30 months depending on site conditions. A single block can be installed as a 400 MW power

6.3: Thailand. Bang Pakong: the second combined cycle there completed 1992, was the first of four plants built by EGAT using a common 310 MW block design.

station in its own right, or several can be installed on one site to share some common services such as the control room, high-voltage substation and water treatment system.

The single-shaft is a more compact and efficient system. In the previous generation of plants, in typically a 2+2+1 arrangement, there would be three 130 MW generators instead of the single 400 MW unit which is marginally more efficient. A larger gas turbine of 250 MW operates at a higher temperature and has a higher efficiency. There is also the potential for heat recovery into the steam cycle.

A further development to improve the reliability and performance of the combined cycle has been the introduction of the once-through, or Benson, boiler. As steam pressures increase so the thickness of the tube walls and the drum has to be increased, so that the time taken to heat the water is longer due to the greater thermal inertia of the drum. With a Benson boiler there is no drum, just a long serpentine tube fed by a circulation pump which establishes the output pressure, and there is consequently a smaller volume of water.

The first use of a once-through boiler was for the repowering of a 120 MW steam turbine at the Rhinehafen power station in Karlsruhe in 1996. It was the first commercial installation of Alstom's 260 MW GT26 gas turbine, which in its initial version had an exhaust flow of 562 kg/s at 610°C. As a result of this, the company were approached by

6.4: Blackstone, MA, USA: American National Power's standard GT24 single shaft block with Benson boilers including SCR units for ultra-low NOx output.

American National Power to design a standard single-shaft power plant for the 60 Hz GT24, which could be a merchant plant in the deregulated American market.

With a merchant plant there is no long term sales contract to a specific customer: instead, power is bid on the spot market 24 hours ahead. Since this depends on there being high availability to exploit the best market opportunities, ANP wanted high part load efficiency and fast starting capability. The result was a 260 MW single-shaft block with a 2-pressure boiler with a once-through high-pressure section at 160 bars, 565°C.

A feature of the GT24 combined cycles is the two-speed steam turbine. To improve the efficiency of small industrial steam turbines, Alstom's Turbomachinery Division, in Finspong, Sweden, who in earlier times had been a major supplier of marine steam turbines, found that by splitting a single cylinder turbine into high pressure and low-pressure stages and changing the rotor diameters and speeds to optimise the efficiency of each section it would result in a more efficient turbine even allowing for the gearbox dropping the high pressure output to synchronous speed.

Known as VAX, among the first schemes in the United States to use this turbine was Richmond Cogeneration a 270 MW combined cycle with two GT11N gas turbines and a 90 MW VAX turbine completed in 1991. Four years later Air Products installed a single-shaft combined cycle for a combined heat and power scheme in Orlando, Florida. This was a single-shaft block with GT11N and a 50 MW VAX steam turbine which went into operation in 1995.

The company were already looking at the combined cycle market for the VAX turbine and in June 1999 installed a 160 MW single-shaft combined cycle at Dighton, MA. This plant had the more powerful GT11N2 and a 50 MW steam turbine. Dighton, and the Orlando unit before it, can be considered to be the test plants preparing for the GT24 single-shaft units to follow.

The two major European gas turbine companies have effectively standardized on the single-shaft block. But some plants have been built to evaluate new principles. Cottam, near Nottingham, England, was the test plant for the Benson heat recovery boiler which Siemens applied to a 400 MW combined cycle based on their 280 MW SGT5 4000 F gas turbine. The boiler was designed to operate at 160 bars with a future, larger gas turbine but has only ever operated at 125 bars.

The G-class was an intermediate stage to improve cooling of the American- and Japanese-designed machines with the can-annular combustor system. Westinghouse announced the W501G in 1995 as an uprated version of their W501F and incorporating steam cooling of the transition ducts connecting the combustor cans to the power turbine inlet. Mitsubishi similarly announced the M501G to be followed by the larger 50Hz model, M701G.

Finally, the H-class has been launched first in the 50 Hz market. The two gas turbines that have so far been designed for this class are both rated at 330 MW and are exclusively for combined cycle application. The target is a nominally 500 MW single-shaft combined cycle block with an efficiency of over 60%.

The first combined cycles with the E-class gas turbines were mainly multi-shaft arrangements with the majority having two gas turbines and one steam turbine. Combined cycles of this period with two gas turbines generally ranged between 300 and 450 MW output with a best efficiency between 48.5 and 52%.

Several were built with three and even four gas turbines, particularly in Southeast Asia. With more than two gas turbines there is an advantage in part-load operation since only one gas turbine could be shut down, over night, leaving the others running with some loss of output and efficiency of the steam turbine, but less than if one of two gas turbines had been shut down.

In Southeast Asia, particularly, combined cycles were built in phases because of the high rate of growth in electricity demand which in some countries reached over 15% per year after 1990. First the gas turbines would be installed and brought into operation, generally within twelve months of the date of order.

These would run for about two years while the heat recovery boilers and steam turbine were erected and then shut down for connection to the boilers and commissioning of them and the steam turbine. The original gas turbine exhaust stacks were retained with a diverter valve in the base so that they could act as a bypass stack and allow the gas turbine to run independently or be shut down for maintenance.

The next generation F-class gas turbines were announced first in the United States in 1989, and were altogether bigger machines with outputs ranging in their current ratings from 150 MW to 280 MW for the 50 Hz market and with exhaust temperatures at around 600°C. Later the scaled down F-class models were introduced, rated at 60 MW at 5200 rev/min, and driving through a gearbox.

Early models of these F-Class gas turbines offered an immediate gain in efficiency from 52 to 55% partly through the increase in gas turbine efficiency, which is now greater than that of the equivalent size steam turbine operating on a subcritical steam cycle. The higher exhaust mass flow and temperature permit use of a tri-pressure steam cycle with reheat. With the introduction of these gas turbines and subsequent upgrading, there was the move to the single shaft-block which, with careful husbandry of heat losses around the cycle and its cooling systems, has pushed efficiency to over 58%, in the 50 Hz market and with the promise of 60% with the new H-class gas turbines.

Also with the larger gas turbines over 280 MW in their 50 Hz models, there was renewed interest in repowering steam plants but for specific reasons. Replace a worn out boiler on a 120 MW set and the power plant, with a refurbished steam turbine matched by reblading and other changes, to the heat output of the gas turbine, can result in a substantial improvement in efficiency at a higher output and move the plant back up the merit order.

In Singapore, at deregulation, Senoko Power, with four power plants at a congested site on the north coast of the island elected to repower the three 120 MW steam sets in their oldest station, dating from 1970, rather than build a new combined cycle. The steam turbines were extensively refurbished and the three fired boilers were each replaced with a GT26 gas turbine and heat recovery boiler. The first set was repowered about three years before the others, so that they could stay in operation until the 260 MW gas turbine had been commissioned. With the installation of the gas turbines the capacity of this first station was effectively trebled.

6.5: Cottam, UK: 400 MW combined cycle with prototype Benson heat recovery boiler. In the background EdF Energy's 2000 MW coal-fired power plant.

In the single-shaft block, the power train of gas turbine, generator, clutch and steam turbine is effectively a standard assembly. Normally it would be erected in its own building with the power train axis about 5 metres off the ground so that auxiliary systems can be mounted underneath. A 400 ton overhead crane runs the length of the building with a roadway running alongside the power train to enable large items of equipment to be picked up and placed in position. Air intake structure and generator connexions are mounted on the other side of the unit so as not to obstruct the roadway, which can serve as a lay down area for large components during a maintenance outage.

Variations would normally be in the cooling system and the heat recovery boiler which may be site-dependent. The boiler could be a drum type or a once-through Benson type, in horizontal or vertical format. The cooling system could be either a wet mechanical draught, or wet/dry hybrid unit, even on a riverside site. The hybrid system has a dry section at the top below the fan, which initially cools the water before it is sprayed onto the packing below to complete cooling. The advantage of this two stage system is that there is no visible plume for much of the year, but if it does occur then it is a much smaller plume.

Alternatively, an air condenser has no visible plume but a larger auxiliary load, since a 400 MW block with a 130 MW steam turbine would have at least twenty four fans. So the auxiliary load is somewhat

6.6: Midlothian, TX, USA: GT24 single-shaft combined cycle block with once through heat recovery boiler. One of six units on the site near Dallas, TX.

greater and the efficiency is lower by about 1 percentage point.

The majority of combined cycle plants ordered in recent years have been in the form of a single-shaft block with the gas turbine and steam turbine driving from opposite ends of a common generator. The gas turbine is connected to the generator at the cold, intake end, and exhausts directly into the heat recovery boiler. At the other end of the line is the low pressure cylinder of the steam turbine which can either exhaust axially to the condenser or in the case of a double-flow cylinder into a side-mounted condenser. The steam turbine rating is about 130 MW in a two cylinder configuration.

The F-class Alstom GT24 and GT26 both have a 32:1 pressure ratio and have two bleeds from the compressor for blade cooling in the power turbine stages. First the hot air streams are passed through two once-through coolers which are connected in parallel on the steam side between the high-pressure economizer and the superheater outlet. In this arrangement it is not the gas turbine fabric which is directly cooled so the once-through coolers always contain liquid even when the gas turbine is shut down.

The G-class gas turbines with steam cooling of stationary components linked the gas turbine more firmly to the steam cycle. They were designed for base-load application at high power and efficiency with the gas turbine coolant taken from the intermediate pressure superheater.

The cold flow from the high pressure cylinder back to the reheater goes first into the transition pieces before it joins the intermediate pressure and goes on to the reheater.

The first H-class gas turbine was GE's response to the Westinghouse G class for the 60 Hz system, but was instead directed at the 50 Hz market. The Frame 9H was announced at the end of 1995 and at 330 MW would have the highest output of any gas turbine produced up to that time. Uniquely the Frame 9H has steam cooling of the first and second vanes and rotating blades of the power turbine. As with the other steam-cooled machines the steam path through the gas turbine is tied into the reheater and intermediate pressure flows.

At the end of 2008 there was only one Frame 9H in operation on a site at Baglan Bay, near Swansey, South Wales, where it has been in commercial operation since September 2003. Baglan Bay is a single-shaft block rated at 480 MW. As a prototype unit it was very much a test bed for the gas turbine which had extensive factory tests of components and systems before it was shipped to the UK, where it arrived on site at the end of 2001. It was fired up for the first time in November 2002. Some nine months of testing the combined cycle followed before the official opening, by the Secretary of State for Wales, in September 2003.

A few months after it entered service the economic environment began to change as the price of oil approached $80/bbl and beyond in response to increased demand from the expanding economies of China and India. The price of gas followed and, for a time, a number of combined cycles were shut down and mothballed until a more favorable price of gas appeared. Here we had the spectacle of some of the newest, most efficient and environmentally friendly power plants being taken out of service because they were not profitable. There was enough spare capacity in an old coal-fired power plant of which capital costs had been paid off years ago to take over the electricity supply until gas prices returned to lower levels.

Changing times have revealed that the inherent operational flexibility of the combined cycle has been overlooked in the rush to install more efficient gas-fired plant for base load capacity. Steam plants were designed in the past for base load or mid and peak load duties, and it is the same with combined cycles.

When the order for the first GT26 was announced it was to repower a set in the Rheinhafen steam plant in Karlsruhe, Germany. However the steam pressure at 156 bars was considered to be too high for a drum-type boiler, and instead a once-through Benson-type boiler was used, which

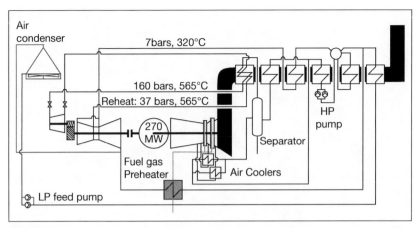

6.7: Schematic of Alstom single-shaft combined cycle for a GT24 Merchant plant with Benson heat recovery boiler and heat recovery from cooling air tor turbine stages.

with no drum and therefore a smaller water volume, can heat up quicker behind the gas turbine.

The design was carried over to the 60 Hz merchant plant based on the GT24 between Alstom and American National Power, who have fourteen units in total with ten on two sites in Texas and two on each of two sites in Massachusets, The design of this single-shaft block features a 2-pressure Benson boiler at 160 bars, 565°C, reheat at 37 bars, 565°C, and a drum type low-pressure stage at 7 bars, 320°C.

Three other New England operators have another eight units between them. These twelve units bid into the New England power pool 24 hours in advance and are noted for high efficiency and rapid start up in little more than an hour. But because of this they can run at peak times when electricity prices are highest. In summer time with high air conditioning loads they could run fourteen hours a day and shut down overnight. At another time it might run several days continuously to cover a nuclear refueling outage or a maintenance outage on a large steam plant.

Siemens as Mark Benson's original licensee have had a boiler research division since 1926 and some fifteen years ago developed a Benson heat recovery boiler for the combined cycle. The unit at Cottam, England, went into operation in 1999. In nine years of operation it has run periods of base-load and at other times in mid- and peak-loads with more than 1200 starts to date.

In 2005 Siemens made the magnanimous gesture of offering to licence the Benson heat recovery boiler to all the manufacturers of heat recovery boilers. So far relatively few have been ordered with four in

6.8: Baglan Bay, South Wales. This 480 MW combined cycle was the first with the GE Frame 9H gas turbine, which went into commercial operation in September 2002.

Germany, two in the UK, one in the Netherlands and eight in Japan by the end of 2008.

The Benson boiler will be exclusively offered with Siemens new H-Class gas turbine, the 330 MW SGT5 8000H which was announced in October 2005 and since January 2008 has been on test at Irsching, preparatory to combined cycle conversion. Like the GE machine it is designed solely for combined cycle use with a 530 MW single shaft block and an efficiency of over 60%.

From this we can deduce that Class H is defined by high power and efficiency of 60% in a combined cycle. The Siemens prototype was shipped from Berlin in April 2007 to the test site in Bavaria. Here there is a 360 MW steam plant of E.ON Kraftwerk, which is being demolished following installation of a new 400 MW F-class combined cycle in 2008. Testing has been completed on the new gas turbine, the design has been validated and the conversion to combined cycle is in progress.

With almost ten years separating the launch of the two H-Class gas turbines the SGT5 8000H is completely different. It is air-cooled and designed for mid-load operation, typically with weekend or overnight shutdown. At the time of the announcement the rapid economic growth in China, India, Brazil, and Russia was pointing to a shift in the pattern of global demand for electricity.

The world population in the year 2000 was 6.1 billion and could

6.9: Irsching, Germany: The prototype SGT5-8000H gas turbine under erection at the test site prior to start of testing in January 2008.
(Photo courtesy of Siemens)

reach 7.5 billion by 2020. In the same period electricity demand is
expected to grow from 15 500 TWh to 27 000 TWh but most of this is
expected to occur in the newly developing industrial countries who are
expected to account for 45% of global demand in 2020 as compared
with 29% in 2000.

Siemens were anticipating an expanding market which would require
larger and more efficient power plants. SGT5-8000H will only be sold
as a combined cycle and will be offered only with a Benson boiler. First,
this will be to achieve rapid starting, and second, with higher steam
conditions to improve efficiency.

The original Benson heat recovery boiler at Cottam has only operated
at 125 bars, which is the normal high pressure output on the current F-
Class combined cycles. With a gas turbine of 17% greater output and a
higher exhaust temperature there is the possibility to go to 160 bars with
the H-Class unit.

As gas turbines have got larger with more powerful steam cycles the
time taken for a combined cycle to run up to full load has increased.
Higher steam pressures have required thicker tubes and drums with
consequently greater thermal inertia. Therefore the gas turbine, when
it starts loading, has to stop at part load while the boiler heats up and
is producing the required steam quality for the turbine. Only then can
loading continue. If the gas turbine is steam-cooled then there is the

additional complication of putting steam into the cooling paths when the gas turbine has reached about 10% part load. Some combined cycles in the United States have been known to take as long as three hours to run up to full load: this with the smaller 60 Hz G-Class gas turbine of 260 MW output. But we have come to a time where the requirement of a combined cycle is for greater operational flexibility throughout its life

Particularly in the United States where combined cycles move off base load in later life, there have been instances of fatigue failures of welds between tubes and drums, on the large horizontal boilers, as a result of more frequent starting.

The Benson boiler is one solution for new plants because it can heat up as fast as the gas turbine and with neither a drum, nor a large volume of water, there is a much lower thermal inertia. The gas turbine still has to wait for the steam turbine to load, but does so at full load when emissions are at their lowest.

Other changes are being made to improve the starting capability of existing plants. Partly it would mean having an auxiliary boiler to stay on overnight to keep pipe work and casings warm. Other changes are in the gas turbine control software, which would make it run up quicker.

In little more than 50 years the combined cycle has appeared on the scene as the ultimate gas turbine application for power generation. The majority of plants around the world are fired with natural gas which is the cleanest of the fossil fuels in terms of its transport, preparation at site and the absence of any solid residue to be disposed of. A few gas fields contain a small percentage of hydrogen sulphide are termed sour gas, which is exclusively supplied on a dedicated line to a power plant, but in general natural gas is a sulphur-free fuel.

Development of gas turbines has been accompanied by the production of clean combustion systems with low levels of nitrogen oxides and carbon monoxide. But the biggest gain of all has been in the efficiency of power generation. Not only is the combined cycle more efficient but the gas turbines driving it are now more efficient than the largest of the subcritical steam turbines.

The challenge now in Europe is to replace capacity which must be shut down in 2015. This is particularly critical in the UK where most of the existing coal-fired and nuclear sets remaining in service will have been taken out of service by 2023.

In April 2009 eight combined cycle plants totaling 10 525 MW were either under construction or awaiting consent with all but the second Drakelow station planned to be in service before the end of 2015. All five plants under construction in 2009 are configured as two or more

TABLE 6.1: UK COMBINED CYCLE PROGRAM IN 2009

Developer	Site	Output MW	Gas Turbines	Service Date
RWE Npower	Staythorpe	1650	4 x GT26	2009
Centrica	Langage	800	2 x GT26	2009
E.ON (UK)	Isle of Grain[1]	1275	3 x GT26	2010
RWE Npower	Pembroke	2000	5 x GT26	2012
EdF Energy	West Burton	1350	3 x Frame 9FA	2013
Scottish Power	Damhead Creek	1000	Awaiting consent	2014
E.ON (UK)	Drakelow	1275	n/a	2015
E.ON (UK)	Drakelow[2]	1275	Awaiting consent	2017
Norsea Pipelines	Seal Sands[1]	800		2011
Total		10625		

[1] Combined heat and power schemes for LNG terminals
[2] Application to double station capacity by 2017

single-shaft blocks of about 400 MW, which with a nominal capacity of 7550 MW, in service by 2015 would more than cover plant closures planned at the end of that year. In any case these old coal-fired plants are restricted in their output until they are shut down.

The Isle of Grain plant is a combined heat and power system which uses hot condensate to evaporate LNG at the neighboring terminal. Consent for a second plant of this type was awarded to Norsea Pipelines Ltd to be built alongside their gas terminal at Seal Sands in Northeast England.

Scottish Power's Damhead Creek site is an existing combined cycle which they took over in 2008. They have applied for consent to build a second plant on the site. It will be the fourth combined cycle on the Isle of Grain to provide replacement capacity for the Kingsnorth coal-fired power station which shuts down under the terms of the LCPD in 2015.

Combined heat and power, has really taken off with the deregulation of electricity supply around the world, and made a significant improvement in the economics of production wherever it has been applied. The gas turbine is particularly suited to these applications with it high exhaust temperature.

This is why gas must be considered as a green option for the future. It has however been used in different ways throughout the world. In the developing countries of Southeast Asia and the Pacific region it was used to support load growth. In Europe and North America it was used partly to support load growth but also to provide economic heat and power to industry. In combined heat and power the even higher efficiency has also contributed to a significant saving in industrial greenhouse gas emissions

Combined cycles when they started were considered to have a life of about 25 years which was more or less confirmed with Korneuburg B in Austria when it broke down in 2004. It had gone into service in 1981. In an unusual turn of events, the original manufacturer, now incorporated in Alstom, had in storage in its Berlin factory a new thermal block for the gas turbine, a GT13D, which had been built against an order that had subsequently been cancelled. The new block replaced the Korneuburg unit, and in addition a low-emissions combustor was fitted, and an upgrade applied to the power turbine after which it returned to service the following year.

The 1835 MW Teesside combined heat and power plant is now jointly owned by GdF Suez which includes Electrabel, who are the current operator. The plan is to replace some of the eight 133 MW gas turbines, which have been in operation since 1993 with four 300 MW gas turbines and replace the two steam turbines with new 340 MW units, effectively making two 940 MW combined cycle blocks. The station output will remain the same but the efficiency will be higher. The same operations will continue with the main power and steam output to the chemical works and electricity sales into the grid.

GdF Suez planned the upgrade to improve the competitivity of the station, while at the same time maintaining basic services to its existing industrial and utility clients. It will be carried out in three phases with the first two covering the construction of the new blocks and the closure and removal of the last of the original gas turbines. The rebuild is expected to be completed by 2014.

Increasingly a number of combined cycles will pass 25 years after 2015 and while some of them can be furnished with upgrade kits to the combustor and power turbine to extend their life at marginally higher output, some will be probably be shut down and replaced with more efficient units.

Many of those built since 1995 have been designed primarily for base load duty at a time when gas was cheap. But their replacements will be required for more varied duty in the future. If we look at the other options for power generation there are none that are as versatile and flexible in operation as the combined cycle.

A nuclear power plant is designed for base load operation and many operate continuously from one refuelling outage to the next, and since a typical refuelling outage takes up to four weeks this is also the time for maintenance of the turbo machinery and auxiliaries. Nuclear plants cannot support large and frequent changes of load, though new designs of gas-cooled reactors may have similar operational flexibility.

Large supercritical coal-fired plants with Benson boilers, are designed for base load operation up to 7000 hours/year. They can follow load and carry a large spinning reserve. They can support seasonal changes in load as for example if they are sending steam to district heating condensers during the winter months. If they are shut down at weekends they can be restarted relatively quickly for the Monday morning peak.

But the continuation of coal even with the integrated gasifier combined cycle (IGCC) concept is problematic and even more so if carbon capture is included. Preliminary IGCC designs have lower efficiency than a large supercritical steam plant and the highest auxiliary load of any power plant yet produced.

Wind energy is only available when the wind blows within defined speed limits and is produced by a large number of small units, not all of which may be operating at any one time. So wind farms must have back-up power which means combined cycles for rapid response.

So much has combined cycle efficiency improved in the last 20 years that some benefit would be obtained by replacing some of the early plants with more modern units. The early plant with the E-class were multi-shaft schemes, the majority with two gas turbines and a steam turbine with output of 350 to 450 MW and efficiency between 48.5 and 52%. The replacement plant could be a single-shaft block of F-class at 400 MW with an efficiency of 58%, or by 2015 there might be the option of an H-class single-shaft block at 530 MW and 60% efficiency.

So a European gas turbine market is beginning to appear outside of Spain and Italy, which with Turkey have accounted for most of the recent combined cycle orders. Some of the new combined cycles will be definitely replacing old power plants, some of which have only recently been demolished.

A number of them are being supplied with Benson boilers for flexibility. Indeed a combined cycle ordered for the Sloe Industrial site in Rotterdam has been ordered with Benson boilers because the owners, a partnership of Delta NV and Electricité de France, decided that the plant will not run overnight because with the current price of gas and the low night load it would not be profitable to run it.

But if a power system of the future is composed of nuclear plants, wind generators and some hydro stations, combined cycles will be required for the mid-load duty and may well operate with daily shut down and no more than 5000 hours operation per year.

One study by Consultants Parsons Brinkerhof, at their Manchester,

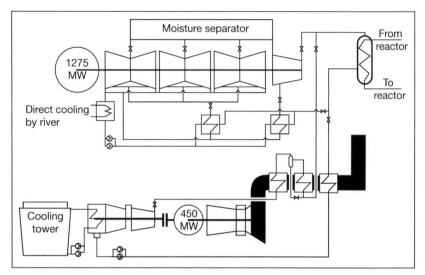

6.10: Integrated reactor and combined cycle proposal to provide extra output by using the combined cycle to provide feedwater heating for the reactor.

UK, division was for the linkage of a combined cycle with a Pressurized Water Reactor; the Integrated Reactor and Combined Cycle (IRCC).

The reactor produces steam at 60 bars with about 0.5% wetness, and preheats feedwater with bleeds from the steam turbine. If the bled steam is sent instead to the combined cycle boiler, it only has to be superheated to generate more power. The combined cycle condensate is then returned to the low-temperature section of the heat recovery boiler which acts as a separate preheater and returns it to the reactor at the required temperature.

The idea is proposed for an existing 1200 MW reactor in the United States which has been relicensed. But the arrangement equally could be made with a new reactor. Both the combined cycle and reactor would be capable of operating separately and switch between separate and combined operation on load.

This arrangement could exploit the flexibility of the combined cycle to meet the different load conditions during the day. The reactor would run continuously on base load and each weekday morning the combined cycle would start up to take over from the nuclear feed heaters and run for up to 14 hours a day before it shuts down and the reactor takes over the feed heating during the night. In other words there is a 1200 MW nuclear generating system which is able to increase output by 33% by sending its feed water to be heated by an external thermal system.

The combined cycle would, of course, be able to run on it own during refueling and maintenance outages of the nuclear plant. But it would not be an ordinary combined cycle. The bled steam volume would determine the capacity of the combined cycle boiler so for an F-class gas turbine such as Siemens 189 MW SGT6 5000F it could be a single pressure Benson boiler with a separate condensate preheater for the same volume of water.

When operating on its own the combined cycle would have to evaporate a part load of steam with the gas turbine exhaust energy. For a normal combined cycle with this gas turbine the steam turbine would be rated about 90 MW: for the nuclear link it would be designed for over 200 MW, and would be running independently at about 35% part load without the nuclear input.

In southern Europe and North Africa, any country in fact with potential for a solar powered steam plant a similar coupling arrangement between the solar boiler and the combined cycle would again increase the output of the combined cycle during daylight hours which is mainly when the combined cycle of the future will be required to operate. One such scheme is under construction in Morocco with a solar steam generator sending part of its output to the combined cycle.

These special combined systems apart, the future combined cycle must be a flexible system to fill the supply gaps caused by maintenance outages and the renewable energy sources.

Rapid starting capability will be essential for the combined cycles of the future. They may not be so fast as some of the early plants with the E-class gas turbines where the steam conditions were lower and boilers heated up quicker. But combined cycles with high-pressure Benson boilers in the United States can come up to load on a warm start in about an hour.

The combined cycle of the future will probably be a single-shaft unit with a Benson heat recovery boiler. Not only does this heat up as fast as the gas turbine to full load but while waiting for the steam turbine to reach synchronous speed the gas turbine will be running at full load with minimum NOx emissions

The higher the power rating of a gas turbine the higher the exhaust volume and temperature and with now at least five models with exhaust temperatures significantly over 600°C there is scope for wider use of the Benson type heat recovery boiler. This will give faster starting capability and with higher steam conditions this will improve combined cycle efficiency. To go significantly above 60% efficiency will require the high steam pressures with a gas turbine able to support them.

But simplicity is also required in the gas turbine designs to ensure ease of maintenance. There may be scope to improve combustor design so that lower emissions are possible at even higher temperatures so that selective catalytic reduction modules would no longer have to be installed in the heat recovery boiler.

From this we can conclude that the combined cycle will be an important component of the green energy system, and not only because of its high efficiency. This suggests that in future to get the best advantage of the combined cycle it should be designed for flexible performance over its entire lifetime. It may be required to run base load for the first few years, but it must be designed so as to be able to continue with the same reliability as a mid-load and peaking unit for the rest of its life.

7

New energy technologies

To bring about a green energy system there are a number of developments under way which must be brought up to commercial application. These are not applications exclusive to one country, but rather designed to bring the benefit of clean efficient energy technology to the whole world.

Of particular concern is the rising cost of fuel and raw materials which of course impact on construction costs. Generating companies want smaller generating plants which are cheaper to build and operate, and this has spawned a number of ideas in nuclear technology to design smaller reactors which can be installed on any power system in the world.

With no emissions and a relatively low fuel cost, nuclear energy has everything going for it. Two small reactors, one gas-cooled, the other a Pressurized Water Reactor are at the forefront of these developments, and are international projects with the aim of commercial operation before 2015.

The Pebble-Bed Modular Reactor under development in South Africa, China and the United States is the one universally applicable system with the potential for a wide range of process applications. It is a relatively small unit of about 160 MW and with the first unit at Koeburg, South Africa, due to come into operation in 2014

Development started in Germany and a 300 MW commercial prototype was built but it only ran for two years before determined Green anti-nuclear protest brought it to a conclusion and forced the abandonment of the technology.

If there was any better proof that the Green movement are nothing but economic saboteurs it is this. Enough experience had been obtained from the 15 MW prototype at Jülich and the two years of operation at Hamm to show that it was intrinsically safe and could be built in a size which could be carried on an electricity supply network anywhere

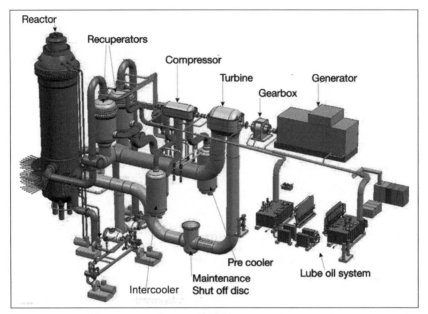

7.1: This model shows the arrangement of the Pebble-Bed reactor as designed for electricity generation with a closed-cycle gas turbine output. (Diagram courtesy of PBMR Ltd)

that could support a 200 MW generating set. It would be the answer to solving the problems of emissions from fuel-fired power stations at a time when environmental clean-up had really started with the concerns over acid rain and a growing market for unleaded gasoline.

Had development continued in Germany the likelihood is that a number of these reactors would now be in service; some even on large islands. Being small and of a standard design they would have a large degree of factory assembly. The generating unit of the new plant is a closed cycle gas turbine which is being developed in Japan by Mitsubishi Heavy Industries.

The development is funded 85% by the South African Industrial Development Corporation, and the state utility, Eskom, who have placed an order for 14 PBMR units following successful operation of the commercial prototype. The remaining 15% is held by Westinghouse who, having taken it over from British Nuclear Fuels, are now seeking certification of the reactor in the United States.

The PBMR as a 160 MW commerical prototype is contained in a steel pressure vessel measuring 27 m x 6.2 m diameter. It has a graphite lining 1 m thick which serves as an outer reflector and this is drilled vertically to accept the control rods. A graphite column on the axis of the pressure vessel acts as an inner reflector, thus creating an annular gas space for the circulating fuel pebbles.

The fuel design is based on the original pebble produced for the German projects by Nukem GmbH, and manufactured in a new factory at Pelindaba, near Pretoria. At full output the factory can produce 270,000 fuel pebbles a year.

Each pebble is a sphere 60 mm in diameter and contains 9 g of uranium dioxide enriched to 10%. which forms a kernel made up of 15000 coated particles which are mixed with graphite and a phenolic resin to form a moulded sphere 50 mm in diameter. To this is added an outer structural layer of carbon, 5 mm thick which is then accurately machined to 60 mm diameter.

Note the much greater enrichment of the fuel to 10% which provides for a greater burn up of fuel which does not stay in a fixed position in the reactor. Each pebble is moving at random as it passes down the vessel. As the pebble leaves the bottom of the reactor, the reactivity is measured to determine the amount of burn-up and if it is not used up it will be returned to the top of the reactor. Each pebble should pass six times through the reactor.

The full fuel load of the reactor is 4.1 t, comprising about 45,000 pebbles. In operation the core is reminiscent of a fluidized bed with the pebbles slowly moving through the reactor from top to bottom over a period of about three months, Over a design life of 40 years the reactor will use some 60 t of fuel.

The helium coolant gas enters the reactor at 91.5 bars, 500°C and leaves at 900°C to enter the turbine which drives the generator at 3000 rev/min through a gearbox. The gas leaves the turbine at 26 bars, 500°C and returns through the recuperator before entering the two compressor stages separated by an intercooler and thence to the reactor.

As designed, more than one of the reactor systems can be mounted on parallel axes like the large steam turbines of the past. They can be installed in the one building at intervals to follow load growth, or they can be installed singly at industrial sites.

There are three versions of the reactor planned. First is the generating plant which is the immediate requirement. This is 160 MW at 41% efficiency and applicable to either system frequency. Construction of the prototype will be complete by 2013 before fuel loading and with load testing to be completed in 2014.

The second version replaces the generating set and in place of the recuperators are two helium/helium high temperature heat exchangers. which are designed as heat sources. The helium output loop provides an extra barrier for the gas loop which is passing through the reactor. and connects to the process heat exchangers

The 160 MW reactor has a thermal output between 400 and 500 MJ/s. A reactor of this size could be used to provide process heat for such applications as coal gasification, hydrogen production and, the distillation of oil from tar sands. All these processes at the moment use coal or natural gas as both fuel and feedstock.

A lower temperature version of the reactor has an output temperature at 700°C which makes it applicable to a wider range of processes. In the case of the electricity generating station, the heat leaving the recuperators and intercooler is similarly available for lower temperature applications such as district heating or desalination.

But all this is in the future, although we have in the PBMR a highly versatile nuclear system which will be available in small unit sizes and applicable both to electricity generation and to a wide range of industrial applications.

The public has to understand that the gas-cooled nuclear reactor is not the same type as at Three Mile Island or Chernobyl; that it is inherently safe, it uses no water, and can be installed much closer to population centres to take advantage of application to industrial combined heat and power schemes and it is completely emission free in operation.

In that situation, too, it would make more efficient use of the existing power system, particularly if it replaced an existing coal-fired power plant. Almost certainly an industrial nuclear plant would be owned and operated by the electricity supplier since they would have the nuclear technicians to operate and perform maintenance, which the client would not have and also provide back-up power to cover reactor outages, as is the case with many industrial gas-turbine power plants today.

The other PBMR application is for power generation on island utilities, such as Cyprus, Mauritius, Sri Lanka, Hawaii and the larger Caribbean Islands These are all places with relatively small networks who have not so far been able to carry a large nuclear unit. With one or two PBMR the utilities would be able to shut down existing thermal plants on the system.The load following capability would also be an advantage in an island network.

PBMR is not the only gas-cooled reactor under development, besides the Chinese project, Westinghouse with PBMR Ltd has a study contract to use the reactor for production of non-carbon hydrogen, which could lead to the application of one of the South African designed reactors in the United States before 2020.

A small water-cooled reactor designed for some of the same markerts

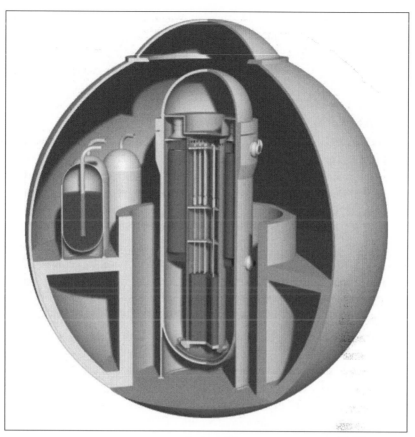

7.2: IRIS Reactor at 300 MW has pressurizer in the vessel head and steam generators arranged around the reactor core inside the vessel.

as the PBMR, but with almost double the generating capacity, is IRIS, a 330 MW Pressurized Water Reactor being developed by an International consortium of twenty countries who are all members of the Global Nuclear Energy Partnership.

The basic reactor design is by Westinghouse and follows on from the AP 600 design which has already received US certification. Where IRIS differs from the other PWR's is in the mounting of the pressurizer, steam generators circulating pumps and control rod drives inside the reactor vessel. The object is to design a smaller reactor which is safe and cheaper to operate and maintain. Of the GNEP countries supporting the project, two, Russia and Lithuania, are looking at combined heat and power applications, with Russia studying desalination and Lithuania district heating.

The reactor is said to incorporate safety by design. So the only large penetrations in the vessel are the feedwater inputs to, and the steam connexions out of the steam generators. The Pressurizer is mounted in the reactor vessel head and the eight steam generators are spaced around the reactor core with a circulating pump mounted at the top of each. The only other penetrations in the reactor vessel are for cable connexions to the control rod drives and the circulation pumps.

Compared with the new AP1000 PWR now in production, which is a loop type reactor, IRIS has a much larger pressurizer in the head of the reactor vessel which, with a total volume of 71m³, is 1.3 times the volume of the external pressurizer of the AP1000. However the IRIS core is only one third the size of that of the larger reactor.

The IRIS reactor vessel is larger than that of the AP1000 to accommodate the steam generator modules, at 6.2 m internal diameter and 21.3 m overall height. But it is mounted in a spherical containment vessel 25 m in diameter. The reactor vessel contains the core 14.2 m high with the control rods and the drive mechanisms located in the space above. The eight steam generators surround this core assembly.

The steam generators are being developed by Ansaldo Energy in Italy. The steam tubes are a spiral array inside the tubular steam generator body and connected at the cold end to the circulating pump and leaving by the hot manifold at the bottom.

Like the PBMR one or more reactors can be installed on a site and the absence of any large penetrations of the reactor vessel for steam generators and the pressurizer, as on the existing PWR designs, and the lower power density greatly simplifies the safeguards for loss of coolant which defines safety by design.

While there can be seen to be specific functions of these small reactors, a wider application of nuclear energy will require large scale power plants to back up the system. While PBMR and CANDU reactors refuel on load the other types do not and are generally shut down once a year for refueling and to do maintenance of the turbine, generator and balance of plant.

The other time when flexible power plant is needed is the non availability of the renewables, particularly in years of low rainfall reducing the availability of water for hydro power, or more frequently when the wind is not blowing.

Combined cycles perform these duties at present and are likely to continue in that role. The other requirement is for peak loading particularly in warm weather with high air-conditioning loads and for sudden events, such as plant outages and television peaks. In recent years, particularly in Europe, a number of combined cycles have been ordered with Benson boilers to enable them to start up faster and with lower emissions.

The Renewable technologies in the background are tidal and geothermal. Geothermal energy has been exploited in many parts of the world where steam vents have indicated the presence of hot strata close to the surface. Examples of these are at the Geysers, in California, north of San Francisco, at Lardarello in Italy where the world's first geothermal power plant was installed in 1911 and at Wairakei, in the North Island of New Zealand.

These New Zealand fields are in the long-extinct Taupo volcanic areas. The country has the largest output of geothermal energy with 654 MW operating and the 220 MW Te Mihi plant which will come into service in 2011 and replace the original 181 MW Wairakei plant, which after 50 years is showing reduced steam pressure.

The geothermal resource in New Zealand comes from rainfall percolating down to hot strata near the surface. This means that after a few years some plants are capped and condensate is reinjected to boost steam production. Nevertheless, geothermal energy is more like conventional power generation in terms of its technology and operational availability. Since the output device is a steam turbine it can generate at synchronous speed and be coupled straight into the grid.

Around the world there are a number of granitic deposits which are at relatively high temperatures at depths of 1 km and more which are adding a new dimension to geothermal power. These can be exploited by shattering the rock to create passages though which water can flow so that water is pumped down to the hot rocks which can extend over several hundred square kilometres.

Some of the hottest rocks are to be found in Australia. There is a major area of granitic rocks extending through northern South Australia, and Southwest Queensland and the Hunter Valley region of western New South Wales. Few people live in the area which is off the national grid system, but the Cooper Basin gas fields, in the centre of the continent, have a large electrical load and are close to some of the hottest rock formations. The first geothermal power plants are therefore being developed in this area.

The hot rock potential of Australia is shown on the map and is estimated to provide enough resource for 10 000 MW of power generation. This is particularly important for the three eastern States which get most of their electricity from a series of large coal-fired power stations, the majority of which are more than 30 years old, but it has to get to market and that means long HVDC transmission lines to link the

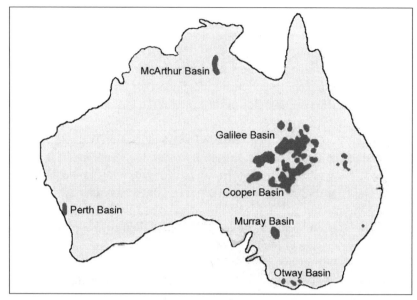

7.3: Australian mainland showing location of Geothermal resources with estimated temperatures above 250°C at 5000 m depth. (Diagram courtesy of Geodynamics)

coastal grids into the remoter areas in the middle of the continent, where the power plants would be located.

There are other small hot-rock deposits near Sydney, Melbourne and Perth, which could be exploited in future. But the main resource is in the centre of the continent with the main development work being concentrated in the Cooper Basin and adjacent areas of western New South Wales.

The Australian company, Geodynamics, has been drilling in the region for the past six years and has established a flow path through a fractured granite at 4250 m depth. The company are now drilling to a 5000 m depth where rock temperatures are up to 300 °C.

In Partnership with Online Energy, Geodynamics has established a circulating loop through two wells at Habanero which are being developed for a 1 MW pilot plant which will take over from diesel generators to power the drilling site and work camp and the nearby community of Innaminka (population 12) who currently spend about $A15 000/year on fuel for a diesel generator.

A second set of wells at Jolioka is going to power a 50 MW generating plant to serve the Cooper Basin gas treatment plant. This separates gas from condensate which is piped down to an export terminal on the coast opposite Adelaide. At present, energy is obtained from gas turbines used

7.4: Soultz sur Foret, France. Water at 200°C is pumped from 5000 m below site to a heat exchanger with isobutane driving the turbine. (Photo courtesy of GEIE EMC)

for compressor and generator drives, some of which will be replaced by the geothermal power plant.

With water coming from up to 5 km depth at up to 300°C and in a closed circuit, a conventional steam turbine cannot be used. The power plant transfers the heat of the water to an organic liquid such as iso-butane which has a boiling point of -11.7°C at atmospheric pressure. This is the working fluid using a low-speed radial turbine. The condensate is cooled in an air condenser so that both heat source and the organic fluid are completely self-contained. The 50 MW Cooper Basin plant is scheduled to come into operation in 2012.

Fractured rock technology is still in its infancy and preliminary work has been to find suitable rock, fracture it by pumping high pressure water down into it and mapping the fracture zone, and then drilling boreholes to receive superheated water to supply a power plant. The most advanced system is in France where in June 2008 a 1.6 MW turbine went into operation at Soultz sur Foret, north of Strasbourg. The heat source is a fractured granite formation 5000 m deep where the temperature is over 200°C. There are two boreholes bringing hot water to the power plant at 50 dm³/s. Cold water returns in a single borehole at 100 dm³/s.

The water passes through a heat exchanger which transfers energy to the working fluid which is iso-butane powering a radial turbine driving a generator at 1500 rev/min. The plant has no waste products and can run continuously at the same or better availability as a nuclear power plant.

Up to now tidal energy has been thought of in classical terminology as a barrage across an estuary with reversible bulb turbines. On the rising tide the water builds up in front of the barrage until there is sufficient head for generation. Similarly on the falling tide the water is held in the estuary until there is again sufficient head to discharge through the turbine.

So there are two periods of generation every day on the rising and falling tides; the times of high and low tide are accurately known and move forward by about 25 minutes per day.

The long term environmental effects of the tidal barrage have deterred planners from looking at this type of scheme. The marine currents which drive the tides are a totally different concept. A tidal current generator is submerged and can also generate at synchronous speed.

Again this is a renewable energy system composed of a large number of small units driving individual generators. The complete system is submerged although the leading design is currently based on a column erected on the sea bed with a control room above water line and provision to bring the generator unit up to the surface for maintenance

At present there are three tidal current generators under development of which Marine Current Turbines have had their 350 kW prototype operating for five years off Lynmouth, North Devon, and have put their commercial prototype 1200 kW system into operation in Strangford Lough, Northern Ireland.

The main test facility for development of tidal and wave power systems is the European Marine Energy Centre (EMEC) off Orkney. This is a research centre which is open to developers to test equipment in the open sea. The tidal site is off the west coast of the island of Eday in an area where there are 4 m/s tidal currents at spring tides. The test site covers an area 4 km x 2km with water depth between 25 and 40 m and has room to accommodate four test rigs.

The Irish company Open Hydro Ltd., has designed a totally submerged system, with a ring of blades rotating inside the outer ring forming the stator of a permanent magnet generator. The water flows through the centre of the device and the energy is captured by the ring of blades. The Open Hydro unit has been installed as a 250 kW prototype in a test rig at EMEC.

The Rotech Tidal generator which is being installed as a 350 kW prototype also at EMEC has won a major order from Korean Midland

7.5: Open Hydro 250 MW prototype on the test rig at the European Marine Energy Centre in Orkney. (Photo Courtesy of Open Hydro)

Power Company for a 300 MW tidal generation plant in water off the south Korean coast. There the system is mounted on a heavy steel base which is deposited on the seabed.

This unit differs from the other units in that the turbine is not directly coupled to the generator. The blades are mounted on a hub which contains an hydraulic pump driven by the rotation of the blades which sends hydraulic fluid up through the hub supports to a generator compartment which contains a 1.2 MW, 11 kV generator driven from either end by hydraulic motors. The hermetically sealed generator compartment is mounted on top of the turbine which forms a removable casette which can be drawn out from the centre of symmetrical venturi ducts and lifted up to a ship for maintenance or replacement.

All tidal generators have the working parts submerged which in turn determines where they can be installed. All systems are designed for water depths of some 40 metres and in the commercial developments are are rated at up to 1500 kW. Two small schemes are proposed for installation off the Welsh coast for commercial operation by 2012.

RWE Renewables are planning a 10.5 MW installation with five units from Marine Current Turbines to be installed at the Skerries, an area off the island of Anglesey about 100 km north of the port of Holyhead. E.ON are planning an 8 MW project using seven of the

Rotec Tidal Turbines to be installed off St David's Head about 30 km northwest of Milford Haven at the entrance to the Bristol Channel.

These are the new energy technologies which are closest to realization. All are free of gaseous emissions, and the geothermal and tidal schemes are already at the prototype stage.

Of the nuclear systems the first small gas-cooled reactors are under construction in South Africa and China, where there is also a 10 MW prototype in operation. Once proven and with the inherent safety of the gas-cooled design this could lead to a wider application of nuclear energy in industrial combined heat and power schemes, which would eliminate the need to burn some of the feedstock to provide the process energy.

8
Renewable uncertainties

Renewable energy must be defined as energy produced by a process which does not require the combustion of a fossil fuel and is taken from a continuous natural source. To date that means a river, the sea, the sun and the wind, and any tree which can be pollarded and produce a fresh crop of wood within two or three years from a sustainable forest.

So from the beginning of human life on earth to about 250 years ago those were mankind's main sources of energy. But there were fewer than 500 million people on the face of the earth and life moved at a slower pace.

The Industrial Revolution from the eighteenth and nineteenth centuries was founded on the ability to mine coal and burn it to generate steam which could then be used to turn machinery from which followed the railways and in their wake the establishment of standard time. Advances in agriculture and medicine brought better diet and greater life expectancy.

The population started to grow and it was some time before birthrates started to fall in Europe. A couple marrying in 1870 would expect to have six or seven children, the majority of whom for the first time might all survive into adult life. Fifty years later in the aftermath of the First World War a couple might have only one or two children.

Since then, international travel, particularly since the arrival of large gas turbine-powered aircraft brought the price down after 1970, has imparted greater knowledge of other countries and other cultures. It has been a two-way traffic, notably between Europe and North America, and international companies have become the industrial giants of the modern world.

The world is certainly energy intensive and yet the efficiency of energy use has improved as demand has increased. Particularly in transport, engines have improved in fuel efficiency and reliability both on land and in the air. For several years it has not been uncommon for three hundred people to fly from Europe to the east coast of the United States and on other journeys of similar length over hostile terrain, in a twin-engined aircraft.

But is this life style sustainable for a world population which has trebled in the past sixty years and is now approaching seven billion? If we depended entirely on renewable energy the answer must be an emphatic no. Even ten years after Kyoto there is no country which can yet produce 10% of its annual electricity supply with wind generators, yet wind generators has been one of the growth sectors in the production of energy equipment over the past decade. Companies making wind turbines have full order books and consequently long waiting times for new orders.

Renewable energy is different from the fuel-based systems that we have used up to now. The availability of hydro power depends ultimately on rain and snow fall, but water can be stored and used when required. Wind depends on the wind blowing which can be at any intensity and at any time of day or night; solar power is only available during the daylight hours and depends on the extent of cloud cover. Tidal power is only available for a few hours either side of high tide. The big difference is that these new renewable energy systems are not continuously available at the time we need them.

It must be remembered too that the development of solar and wind power systems, started not after Kyoto but after the 1973 oil crisis. At that time the electricity supply industry was still the fully integrated, largely publicly-owned power system which had a professional duty to provide for the security of electricity supply across its service area. So the view of the time was to build a large power station to work with the new energy source. But in hindsight, was this the right way to do it?

The big difference with the new renewables beside their availability is their small unit size. The first wind generators were rated a few hundred kilowatts and it has taken 30 years to develop a 6 MW unit which is primarily designed for offshore installation.

Similarly, with solar power the first concept was to have large groups of tracking mirrors following the sun and focusing on a boiler mounted on top of a tower. Plants of this type have been built in the United States and Spain. But it is only in the last ten years that large photovoltaic cell arrays have become possible, so that one tailored to the energy demand of a large building, or even a private house, can be installed on the roof. In effect this is bringing electricity production down to the individual consumer.

The question we must therefore ask is whether, in our desire to prove that renewable energy was possible, have we been doing it in the wrong way? It has taken a long time to achieve any lasting market penetration and we have had to wait for the deregulation of electricity supply to separate generation from transmission and distribution to give it momentum. Once it became possible for anyone to generate electricity then it was possible for an industry to sell surplus power back to the public system when it was available.

Deregulation opened the door first to industrial combined heat and power. Companies who wanted to generate their own power did so because they wanted to cut their energy costs and generally decided to build because they had old boiler plant burning coal or oil that needed to be replaced, and gas was the cheapest available fuel; besides which combined heat and power was more efficient.

Generally the size of the gas turbine was determined by the steam demand. As long as it could produce all of the process steam demand from its exhaust gas energy, the electrical output would be either more or less than their electricity demand so that they could sell a surplus of a few hundred kW to the grid, or else make up the difference from it. This would also leave the security of the grid connection to cover maintenance or repair of their generating plant.

As a working arrangement this is easily contained within the existing system operations. The company can tell the grid when it will shut down its plant and can also forecast changes in output arising from changes in production or the introduction of new process plant. A network could have twenty or more combined heat and power plants on its system.

Now consider a renewable energy scheme. The company has installed not a gas turbine but a wind generator; a single 2 MW unit to supply manufacturing processes which do not require steam, and has machining processes, and a large office computer load.

The availability of wind determines when thecompany will not have to buy power from the grid and on average over the year about 30% of their electricity demand on 260 working days is met from the wind turbine. But that does not mean power is available all day on 83 working days. During some nights and on weekends the wind generator might be operating and its output then can be sold to the grid as can any power surplus to the company's requirements, but during a normal working day it might be that the wind was not blowing, the generator was idle, and power had to be bought from the grid.

The company therefore saves all or part of its electricity costs during some working days, and earns money to set against the cost of electricity bought from their regular supplier when the wind is not blowing. If the wind picks up in the night the payment they will get is less that they would for the same amount of power sold during the daytime according to the tariff structure.

Solar energy is more predictable, since it is only available at a time when the electricity demand is high. An office opens at 8.30 am, two hours after sunrise, and as people arrive computers are switched on, some lights are turned on, and other office equipment and some of the electricity consumed will be be taken from solar photovoltaic panels mounted on the roof. At weekends any solar energy power not used in the building could be sold to the grid.

Perhaps this is not an office but an hotel which has an entirely different pattern of demand depending on how many guests it has, how many people visit for lunch at its restaurant, and how many lights are on in rooms and corridors. Again solar panels on the roof provide some or all of the electricity demand but only in daylight hours and in varying amounts.

Semiconductor developments, particularly of photovoltaic cells, have given a new lease of life to solar energy. In as far as the time of sunrise and sunset are known and that the maximum demand for electricity occurs when the sun is out, it is predictable. The variation between winter and summer daylight hours decreases as one approaches the equator, but it is not only countries such Spain, Italy, Israel and Australia that can easily exploit solar power. Photovoltaics have opened the market in many more countries.

HSBC, the international bank, has at Canary Wharf, headquarters in east London, installed a solar energy array on the roof, where 441 photovoltaic panels have been erected 213 m above datum. The building is not the tallest on the site but is not shaded by the neighbouring buildings. Output is expected to be 77 500 kWh per year

This is not their only photovoltaic system. The HSBC Management training centre near Elstree, 28 km northwest of Canary Wharf has both roof mounted panels and a ground-mounted array which supply 10,000, and 20 000 kWh/year respectively, and have also supplied installations to their First Direct Internet Bank headquarters in Leeds and to a new branch in Attard, Malta, where they have two large tracking panels on the roof which supply 4200 kWh/ year.

Photovoltaic installations have started with large business premises. But how long will it be before smaller installations are possible serving individual houses? When electricity supply was first deregulated the freedom to choose your electricity supplier was gradually introduced,

8.1: London, UK: HSBC has installed this photovoltaic array on their headquarters' roof which is expected to generate 77 500 kWh per year. (Photo courtesy of HSBC)

with the first being companies with high electricity demand, and it was some time before individual households were given the right to change. In fact, relatively few changed their supplier, and of those that did many have since stayed with their new supplier.

The home environment is perhaps the most interesting because in many ways historic developments have paved the way for significant environmental improvements. Houses with electric heating use night storage heaters: a stack of blocks of high thermal capacity arranged around a heating element and contained in a casing forming a duct. The heater stands off the floor and the room is heated by convection with the air flow rising through the casing and out through an adjustable grill at the top.

The concept was devised more than forty years ago in Europe to provide a night-time load for base loaded plants, and was accompanied by a big promotion of night storage heating. Electricity is supplied on a two part tariff with, in the UK, a night-time rate from midnight to 7.00 am when the heaters are charged, and priced at about 33% of the daytime rate.

Householders with this tariff can further exploit it. Modern washing machines and dishwashers are equipped with a single cold water feed which is heated in the appliance to the temperature set for the washing program. Say that the householder has entertained four people to dinner so that there is a large number of dishes, glasses and cutlery to wash up afterwards. If the dishwasher is loaded and the wash program set to start after midnight it will do the job at a third of the cost in electricity.

The householder will see a result in lower electricity costs, but everybody with this tariff, by doing this, is unwittingly helping their electric utility to operate more efficiently by keeping more plants running continuously instead of running down to say 50% part load and lower efficiency every night and running up again the next morning.

If one million houses are fitted with 4 kW solar packages it would amount to 4000 MW of solar power at an availability of about 50% over the year. But it would require a specific law to be put into effect so that the conditions under which solar power can be applied are clearly defined as part of a national programme. For a house with electric heating it would allow water to be heated during the daytime on the solar input and not just rely on a volume heated for seven hours overnight.

Furthermore the solar panels and their supporting equipment must be manufactured and tested and staff must be recruited and trained to install and maintain them, but this is equally true of the other renewable systems that are required in any large number.

The question now is not so much can we generate our own electricity, but is the electricity supply system organised to handle a large number of individual households with varying amounts of electricity for sale up to about 4 kW when the sun is out? If a Government has a policy to achieve a given percentage of electricity from renewables it should be producing the necessary legislation to enable it to happen. If this were the case, there might be more renewable energy in service by now.

Although the Scottish National Party has nailed its colours to the anti-nuclear mast it may be out of power by the time it comes to close the two nuclear stations of Hunterston B and Torness. But for all that, as the present regional government they have started a green energy campaign for small businesses and households with grants of up to £4000 or 30% of the cost to install renewable systems, which have been predominantly solar photovoltaic schemes, and ground-based heat pumps.

Scotland, being on the same latitude as southern Sweden, it enjoys much longer daylight hours in summer than London, and solar photovoltaics will be at their most productive during the summer tourist period when electricity demand in the Highlands is high. But given that the Scottish government can set up a program for renewable microgeneration, why cannot other Regional and National Governments set up something similar?

Intelligent meters are gradually being introduced which will indicate the amount of electricity being drawn at any given moment and what it would cost. If these meters can also give negative readings they would show how much energy is being exported into the grid and define a net cost for power.

There are other renewables which are more conventional in their availability. Biomass (agricultural and forestry wastes) is one that can be burnt either in an existing power plant or in one constructed for the purpose.

There are a number of power stations burning wood chips, municipal solid refuse, and farm wastes such as chicken litter, but these are relatively small. typically up to about 25 MW and located near to the source of the fuel though some with large industrial suppliers of waste can be much bigger. This is what we normally understand by biomass. Since the fuel has to be taken by road to the power plant this limits the economic radius for fuel supply. So biomass has tended to be fuel for specific industries.

Bagasse, as the residue of sugar cane has for long been used as fuel to supply power and steam to the sugar refineries. The power plant would be sited near to the sugar refinery.

Similarly the pulp and paper industry uses the scrap material of the trees, roots, twigs, and bark, and the black liquor as fuel to generate steam for stock preparation and drying of the paper. Combined heat and power plants at mills in Sweden and Finland sell surplus power to the public grid and surplus heat to the local district heating networks.

The 44 MW Stevens Croft powerplant of E.ON (UK) at Lockerbie, in southern Scotland is burning wood waste from sawmills and forestry operations in the area and at the end of 2008 was due to take the first consignment of coppice willow from local farmers who are growing it. In total it needs 480 000 tons of wood per year to produce its full output and the ultimate aim is to have approximately 20% of the fuel coming from coppiced willow.

This is E.ON's first biomass plant in the UK. They have a planning application out for a second plant at Blackburn Meadows, the site of an old coal-fired power station in Sheffield. Like the Scottish plant it will burn sawmill and other wood waste and coppiced willow, and miscanthus grown by nearby farmers.

Miscanthus (elephant grass) is best known as an ornamental grass plant which is available singly from garden centres across Europe. It is grown from rhyzomes which once planted sprout up year after year, and is derived from the temperate variety, *Miscanthus Sinensis* which can be grown commercially in Europe. Once planted it just has to be harvested every year. It requires no special equipment to harvest it and any arable

8.2: Lockerbie, UK: This 44 MW biomass-fired power plant burns wood waste from local saw mills and forestry operations. (Photo courtesy of E.ON (UK))

farmer can plant a few hectares as several have done. The problem now is that a price of over £100/ton for wheat makes it profitable to grow traditional grains for food, rather than specific energy crops. This may not mean a reduction in output, but that no new plantings are made.

Energy crops are not new. It was the 1973 oil crisis which set Brazil producing ethanol from sugar cane, and for more than thirty years cars on the streets of Brazil have been fuelled with a mixture of 95% gasoline and 5% of added ethanol.

To go into the cultivation of energy crops there has to be the land available to grow them, and it must be profitable for the farmer to produce them. Clearly it was in the UK four years ago but not now. The growth of particularly the Chinese, and Indian, economies has pushed up the price of raw materials and food which has resulted from higher prices in Europe and North America for the traditional crops such as wheat, barley and oil-seed rape.

Another factor which has increased crop prices is the use of large areas of wheat, particularly in the United States converted to produce ethanol which, as in Brazil, is aimed at reducing fuel imports; so grain exports have reduced as more is used for bio-ethanol production.

Gas created by anaerobic digestion of refuse in land-fill sites is another energy source and one of the largest such schemes in Europe is in the UK near Solihull. Almost all municipal refuse in the UK goes to landfill and some sites have been organised to collect the gas released and use it to generate electricity. At Birmingham the landfill was extended above ground in clay cells and sufficient gas is collected to drive a 3.5 MW gas turbine which sends its output to Birmingham International Airport. Most schemes are much smaller and use reciprocating gas engines.

In Europe a lot of the refuse is incinerated to generate steam for a turbine. Of course the Green movement don't like it and any plan to extend one of the existing 20 plants or build another in the United Kingdom meets with howls of protest from the Green Activists founded on ignorance. Up to fifty incinerators are planned in the UK which has one of the smallest components of incineration in its waste management policy of any country.

In September 2008 the British Government gave its consent to a 100 MW incinerator to be built at Runcorn, in northwest England. The fuel will be domestic and municipal refuse collected from Manchester, Liverpool and north Cheshire. The plant is being designed as a combined heat and power scheme which will supply process steam to the INEOS Chlor Vinyl chemical works.

The alternatives of recycling and continued landfill do nothing for the environment. Every day people will go to the Local Authority recycling centre taking bottles, and waste paper, which concentrates some of the recyclable materials, and even garden waste to be composted. Apart from all the fuel consumed in going there why do people take garden waste? Even a small plot can hold a compost heap to receive lawn and hedge trimmings dead plants and vegetable stalks and leaves removed from the kitchen.

Spent cooking oil from restaurants and food processing plants and fatty wastes from abbatoirs also can be collected and distilled to produce bio-diesel fuel. These are relatively large sources of material which have to be disposed of anyway. Bio-diesel is used at about 5% in regular diesel fuel, although tests are being carried out with vehicles burning 100% bio-fuel to see what changes it will make to engine wear and performance.

Of all the non-fossil options for generating electricity the biomass option is potentially the largest. with the use of waste materials, from forest and farm and solid municipal waste. In fact in the United States almost 4000 MW is generated by refuse incinerators and landfill gas sites around the country. In Denmark, steam turbines at the refuse incinerators are connected into the local district heating systems.

The new renewables such as solar and wind are in smaller individual units but large numbers of units are being incorporated into larger installations. For example a 100 MW wind farm could have between 30 and 80 individual wind generators depending on the size of the unit and whether it is to be installed on shore or offshore

But the problem with the new renewables is their lack of continuous availability. For a start, if the rotor diameter for a wind generator is more than 120 m, then each unit must be separated from its neighbor by at least 500 m to avoid interference between adjacent units. Thus a 180 MW offshore wind farm with five rows of ten generators would cover a sea area of 67,500 ha whereas a 180 MW combined cycle with two 60 MW gas turbines and a steam turbine would occupy about 4 ha. Furthermore the combined cycle would have annual availability of more than 90% compared with little more than 30% for the wind farm. To match the combined cycle availability of 90% would require forty-five of the fifty units of the wind farm to be running at their maximum continuous rating for 24 hours a day, and 365 days a year.

Also, in the case of the combined cycle the performance is predictable. Planners will know every day that there is gas going to the station to provide generation for the next 24 hours. But there cannot be the same certainty with wind. The weather forecast will tell them if there will be wind, but cannot be precise as to the time and the strength and duration of the wind so that they are able to deduce how many generators could operate at what time of day.

So if renewable energy has a long term future it has to be used in conjunction with the fuel based energy systems of the present. In simple terms it means using use the wind generator when the wind is blowing and solar energy during the daytime. For this reason enthusiasm for wind is greatest in countries where the governments have rejected, or are openly hostile to nuclear power: In Europe that is principally Denmark, Spain, and Germany.

When most people talk of renewable energy thay tend to thimk of wind power. It is the first to have been deployed on a large scale but the largest units in service are currently at about 3.6 MW. An example of a typical European offshore wind farm is at Burbo Bank in the Irish Sea off the north coast of Wales with twenty-five 3.6 MW units, and completed in 2007.

The output is proportional to the square of the rotor diameter so that there is a mechanical limit on the size from the viewpoint of the mechanical loading of a generator mounted on a pivot on top of a tall mast. The wind speed varies and therefore the output is not at a constant frequency, and has to be converted to DC and then reinverted to 50 Hz for transmission.

Public opinion has not been welcoming of wind power since it has to be mounted on high ground to catch the wind or in remote areas which people living there want to preserve for their remoteness. On April 21 2008 planning consent for a 670 MW wind farm on the Isle of Lewis, in the Outer Hebrides was rejected. It had been planned to install 186 x 3.6 MW units but was rejected on environmental grounds.

Four weeks later Shell withdrew from the London Array a larger offshore wind farm of 1000 MW along either side of the Thames Estuary, here the reason given was that rising costs of raw materials and construction made the economics marginal.

A typical offshore wind farm is in shallow areas off shore, as for example, the Irish Sea, the southern North Sea and the Baltic, but the aim is to move out into deeper water on the continental shelf. In fact, anywhere that a drilling platform could safely be installed would be suitable for a wind farm but would it be practical?

The largest wind turbine yet in operation is the Model 5M, developed in Germany by RE Power Systems AG, of Hamburg. Rated 5 MW, it has been designed for installation offshore with a rotor diameter of 126 m. Two of these have been installed in 45 metres of water some 25 km offshore in the Moray Firth off the Scottish coast.

This is a demonstration project which is a joint venture of Talisman Energy(UK) Ltd and Scottish and Southern Energy. The aim of the project is to study the operation of a deep water wind farm and is located close by Talisman Energy's Beatrice oil field. The two generators will supply their output to the platform. The project comes under the European DOWNViND program (Distant Offshore Windfarm with No Visual Imapact in Deep Water). The demonstration will last five years but the aim is to run to the end of their design life of twenty years.

The prototype unit was installed at the RE Power test site beside the nuclear power station at Brunsbuttel on the Elbe estuary. But the unit has been designed specifically as an offshore unit. As can be seen in the photograph, the column is mounted on a tubular jacket support similar to that of an oil platform. The reinforced glass fiber blades are 61.3 m long and the hub height above water level is 85 m. Being in deep water access is not by boat but by Helicopter to a landing deck mounted on the back of the nacelle.

The erection of the units at sea was the first time that a fully assembled generator nacelle, complete with blades and mounted on its column was erected on shore and transported to site in one piece to be lifted on to the jacket that had been built to receive it.

This will be the normal way of installing the 5M since to otherwise erect a large wind turbine out at sea would mean taking the various

8.3: One of twenty-five units of Siemens' largest wind generator, rated 3.6 MW forming the Burbo Bank wind farm in Liverpool Bay. (Photo courtesy of Siemens)

parts out to the site and erecting them from a ship moored alongside whenever sea conditions allowed. With fabrication on shore a single lift is required onto the jacket which is a stable structure standing on the seabed. However once this is done and secured, commissioning teams can be sent by helicopter and all their work will be under cover inside the turbine nacelle.

So far, 43 of the 5MW turbines have been ordered, all but four of them for installation offshore. Of these, the two in the Moray Firth are the first in deepwater and this is interesting because of the possibility of extending the range of sites for offshore wind farms into deeper water where the construction of the jacket to support the unit uses the same techniques as for a drilling platform. It is even suggested that when the oil fields shut down, there will be a use for the structures to carry wind generators.

If classic hydro power is considered renewable energy then not only is there a lot of it still which can be exploited, there are also many sites developed more than 80 years ago which have been abandoned and are now candidates for hydro redevelopment. There are still also areas of the world which depend on hydro power for more than 50% of their electricity supply.

The description of Africa as the Dark Continent first appeared

8.4: Two 5MW wind generators off the Scottish cost supply all their output to the Talisman oil platform shown at right. (Photo courtesy of RE Power Systems)

in literature at the end of the 19th century, but it is certainly true for different reasons today. Fly over Africa at night and there are few places that are lit up as would be any European, Asian or American city.

Africa has the largest unexploited hydro power potential in the world, and half a billion people with no access to electricity supply and dependent on biomass (wood and animal dung) for their energy supply. There is the technology available to build the power plants and transmit their output over long distances but a lack of skilled engineers to maintain the systems in all but a few places.

But any new hydro development in Africa or elsewhere immediately attracts Green protest on the grounds of environmental damage, the obstruction of fish migration and the prevention of silt flowing down to the lower reaches to fertilise the land.

While it may be true that this has happened on some rivers the fact is that historically, hydro power has been developed to improve irrigation and prevent flooding of downstream communities. These are real benefits and the development of more hydro capacity would facilitate rural electrification in parts of Africa so that people could have a clean source of energy for heating, lighting and cooking and not have to rely on the use of crude biomass for their basic energy needs.

The reality of the present situation is that millions of people without electricity in rural areas of the tropics, and not just in Africa, are one of the biggest sources of pollution through their dependence on wood and dung as fuel to meet basic daily energy needs.

The Zaire River is the largest in Africa with a number of large tributaries. What makes it unusual is that the drop in the river level is concentrated near the mouth with a long series of rapids above Kinshasa. The entire river system is estimated to have about 100 000 MW of hydro power potential, of which some 44% is concentrated 90 km from the mouth of the river where it falls 100 metres in the space of 15 km. Two power plants were built at Inga in the 1970's with a total output of 1700 MW to supply the copper mines in the east of the country, but are now not running at full output for lack of maintenance.

The Democratic Republic of Congo has been racked by civil war as have many other countries in the region in much of the last 25 years and it is only now that the conflicts have died down and people are looking again at first, a third Inga power plant, but also Grand Inga which early in 2008 was the subject of a conference in London between bankers and utilities.

Many African countries are seeing big export contracts from China for natural resources of fuel and mineral ores. But while more of this money may be legitimately invested, the Grand Inga project will be twice as big as China's Three Gorges and would have 60 x700 MW turbines for a total capacity of 42 000 MW. It is estimated that it would cost $40 billion to build and the plant could not be completed until 2022 at the earliest.

But where would the energy go? There is nowhere near the demand in Congo for the 372 TWh/year that the plant would produce. There are suggestions that high-voltage direct current links could transmit it as far as South Africa and even to Europe and the Middle East. Given an effective global carbon trading scheme we might see big coal- and gas-fired utilities in Europe buying pollution credits from the African owners of this and other large hydro plants that don't need them. Then again Europe might feel uneasy about a large power transmission line crossing over the Islamic countries of North Africa.

There are also rivers which cannot be exploited for serious commercial reasons, particularly on parts of the Atlantic coast of Europe and the Pacific coast of North America because these are the breeding grounds for the Atlantic and Pacific salmon.

Construction of classic hydro plant takes several years. The river flow and its variation over the year determines the height of the dam and the size of the generating plant. and the first generator cannot be tested until the reservoir is sufficiently full. The only standardization is in the individual generators of the power station, but hydro power, has seen its role in electricity supply changed with the introduction of thermal and nuclear capacity. Its value is in its much greater flexibility of operation.

Small hydro schemes are a totally different system. In the early years of the 20th century many small sites were developed to provide electricity to the surrounding communities. There are others which could have been developed in earlier times but were overtaken by events. Many of these old sites are now being rebuilt and some others are being developed. These are relatively small schemes, generally run of river plants of less than 5 MW, but as green energy schemes earn a premium price for their output. .

RWE through their subsidiary Npower Renewables in the United Kingdom has redeveloped a number of small hydro sites. At present they have 15 plants with a total output 58.8 MW. Typical of these is Blantyre, some 15 km above Glasgow on the Clyde at an existing weir. A single Kaplan turbine generates some 575 kW and came into operation in 1995.

The company are constructing another four stations in the North of Scotland with a total of 10.75 MW. A fifth scheme at Romney Weir on the Thames near Windsor will have hydrodynamic turbines (Archimedes screws) installed in two bays of the weir which will produce about 1.4 GWh per year and will supply Windsor Castle. The scheme which could be similarly applied elsewhere on the river has a net capacity of 300 kW depending on the river flow which could be between 5 and 20 m³/s.

Romney Lock with a drop of 1.46 m is not the deepest on the Thames; further upstream, Boulter's Lock at Maidenhead is the deepest at 2.39 m and above there, Marlow Lock is 2.16 m. Eight years ago serious flooding around Maidenhead led to the construction of the Jubilee river an 11 km flood relief channel from above Maidenhead to reenter the river at Potts Island below Romney Weir, which was completed in 2002.

Deregulation of Electricity Supply in the UK created some opportunities for small hydro particularly in the southwest where a number of rivers flow off Dartmoor, and Exmoor. Several units have been installed to provide energy for companies or individual farmers and landowners either completely or with some export to the grid.

Pumped storage is a variation of hydro power which was introduced with the growth of nuclear power in the 1960's. With this arrangement water is pumped up from an existing lake to a reservoir constructed on the top of a nearby hill. The turbomachinery is a reversible pump

8.5: Blantyre, Scotland: RWE Npower's 575 kW hydro station 15 km above Glasgow on the Clyde was built at an existing weir. (Photo courtesy of Npower Renewables)

turbine so that when water is pumped to the upper reservoir it acts as a pump driven by an electric motor. and to drain the reservoir the pump is reversed and runs as a turbine driving a generator. One of the first was built at Blaenau Ffestiniog, in North Wales, another was built at Vianden, Luxemburg but the largest in Europe so far, with six 330 MW reversible pump turbines is also in North Wales at Dinorwig.

This was completed in 1984 and is entirely underground. The upper reservoir on Llyn Beris was recently enlarged. With the build up of nuclear power in a predominantly thermal system the station was built to carry the loss of two 660 MW sets dropping out simultaneously. All fourteen of the Advanced Gas-cooled Reactors had 660 MW turbogenerator sets, and besides these there were also six coal-fired units at Drax, and three oil-fired units at Littlebrook. So Dinorwig was built for system security with each of the six turbines able to run up to full load in 16 seconds.

This is the classic example of renewable energy coupled to a thermal power system. It was introduced as nuclear power plants came into service, in order to provide them with a night load. This is because nuclear plants cannot follow load to any great extent less it poisons the reactor. So instead of dropping 300 MW at night the reactor keeps running and pumps water up to the top reservoir. Then in the following morning as electricity demand increases the reservoir starts to drain like

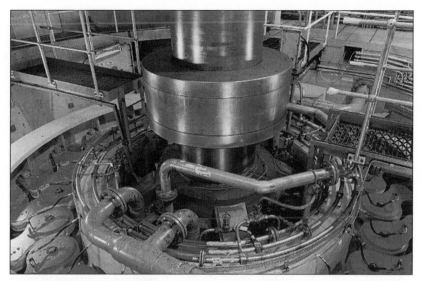

8.6: Dinorwig, UK: one of six 330 MW reversible pump turbines installed for system securit, and all in operation since 1984. (Photo courtesy of International Power)

a conventional hydro plant. It starts as soon as the head gate is lifted, and each unit can supply 330 MW of peak power for as long as it is needed.

In another mode of operation, if a big thermal power plant trips out through a breakdown, the hydro plant may not be able to replace the lost capacity but it could run for, say, an hour while some gas turbines or a combined cycle start up. The point is that the hydro power generator is a fast acting plant which can respond immediately to a loss of generation capacity. It is the only plant that can do this, provided that there is water in the reservoir.

Generally hydro stations operate when required, but it all depends on rainfall, which can vary from year to year. So there can be years when the rainfall has been lower than average and there has been less hydro power generated as a result. In years of above average rainfall the hydro power output is much higher.

Hydro power has been available for a long time and many of the largest schemes are more than 50 years old. There are still a large number of rivers in Asia and Africa which have not been developed because there has not been the demand for the energy within reasonable range of the site.

The latest of these large dams to come into operation is the Three Gorges scheme on the Yangste in central China. The Yangtse, together with the Nile and the Amazon are the three longest rivers in the world. The Aswan High Dam in Southern Egypt is one of only two on the Egyptian stretch of the Nile. There are no dams on the Amazon although a dam was build in the 1980's on the Tocatins the last major southern tributary before the delta and two more are under construction, In total more than 80% of electricity supply in Brazil is from hydro power.

Of the countries with the largest hydro capacity, there are five of which the United States and Canada started first with Brazil, Russia and China following. In Europe much of the hydro power potential was exploited before 1950, except in Scotland where the North of Scotland Hydro Electric Board created in 1943 before the post war nationalization of Electricity Supply was specifically created to develop the hydro potential of the Scottish Highlands.

Sixty years later as Scottish and Southern plc it is one of the six major generating companies in the UK with a large customer base in England and two combined cycle plants at Keadby, near Lincoln and at Seabank south of Bristol. After a gap of more than 50 years the company has built a 100 MW hydro plant at Glendoe which discharges into the southwestern corner of Loch Ness and was completed in early 2009.

When we think of tidal power it suggests a large power station built across a river estuary with reversible bulb turbines to be able to generate on the rising and falling tide. That is the format of the tidal power plant across the estuary of the Rance river in northwest France. There are only three stations of this type that have been built, of which the first, and by far the largest, was the 240 MW station in the Rance estuary, which dates from 1967. There are twenty-four 10 MW reversible bulb turbines which generate on the rising and falling tide. Average output over the year is 68 MW with a production of 595.7 GWh.

The highest tidal range in the world at 16 m is in the Bay of Funday which is the site of a 20 MW station of this type incorporated into a causeway across the mouth of the Annapolis River as it enters the Minas Basin on the Canadian side of the Bay. There are also several river estuaries in Maine and New Brunswick which have large tidal currents and are being considered for other schemes.

Tidal estuaries that could support a power plant similar to La Rance are relatively few and have drawn public objection because of the potential damage to trade and wild life habitat around the estuary.

The Severn Barrage was originally conceived as a 50 km long dam but would cut off the ports of Newport and Avonmouth and limit the size of vessel that could travel through the locks installed for that purpose. But the modified action of the tides would interfere with the wildlife of the estuary which is an important winter feeding ground for arctic sea birds. To build it would cost as much as ten 1600 MW nuclear power stations to provide 5% of the national electricity demand, whereas the nuclear plants would supply over 40% and be available 24 hours a day.

However there are many river and coastal passages with high current flows which are potential sites for a tidal current generator. This device can be likened to an underwater wind generator. The first such device is a 300 kW unit installed in 2003 at Lynmouth on the North Devon coast using a single generating unit working off a tidal current at the entrance to the Bristol Channel.

In March 2008 a larger unit of 1.2 MW was installed in Strangford Lough a large sea inlet in Northern Ireland some 30 km southeast of Belfast. The narrow entrance to the lough results in a fast current with a high power density. This is the first commercial example of a tidal stream generator which went into operation in June 2008.

Developed by Marine Current Turbines (MCT) it consists of a tall column erected on the seabed with two sets of turbine blades turning at about 10 rev/min each driving a generator through gearbox which are contained in nacelles at opposite ends of a beam. For maintenance the whole assembly can be jacked up above water, removed and replaced with a new or upgraded unit. The present unit is designed for installation in water depths between 20 and 40 metres. But the big advantage of tidal energy is that it can generate at synchronous speed.

The site in the narrows leading into Strangford Lough has a maximum current velocity of 4.5 m/s. By adjusting the turbine blade pitch the output of 1.2 MW can be held down to 2.4 m/s on flow and ebb tides. Output could be up to 4500 MWh/year at an availability between 40 and 45%. The unit was installed in April 2008 and went into commercial operation in June the same year. As to when the power will be available, the times of high tide are known and vary within each month. Spring tide, when the Moon and Sun are acting together in their gravitational pull on the water, can equally be forecast.

Now that there are two operating examples of the system in the seas around the UK, things are beginning to move in tidal power, not in the exploitation of the height of the tide, but of the marine currents which drive it. These can be measured in a number of places around any coast to determine the energy density of the flow. There are now at least three other companies with designs for tidal current generators but only one other with an actual unit in commercial operation. The Irish company, Open Hydro Ltd., has installed its 350 kW prototype system at the European Marine Energy Centre (EMEC), in Orkney.

8,7 La Rance, France: 240 MW tidal power station generates on the rising and falling tides and is the largest of its type in the world. (Photo courtesy of Electricité de France)

The Open Hydro design has a rotor consisting of a ring of blades surrounded by a winding excited by a permanent magnet. The whole assembly rotates within the stator ring and the whole is mounted on a heavy steel base which is stood on the sea bed. It is completely submerged. The design is unusual in having no solid rotor axis. The stator is in the form of a ring with the current passing through it. Thus only the periphery of the stream passing through the blades produces the rotation to generate the energy

Another fully submerged design is based on the Rotech Tidal Turbine in which the active flow goes through a bi-directional turbine between the inlet and outlet venturi ducts. The turbine is a conventional hub design supported from the duct section. The rotating blades drive an hydraulic pump inside the hub which sends pressurized oil up to the generator nacelle on top. Hydraulic motors drive a standard 1MW, 11 kV generator at 1500 rev/min and from here the power is sent to shore.

The complete system is produced by Rotech Engineering in Aberdeen for Lunar Energy who are the investment and marketing company. A 100 kW prototype is being installed in 42 m of water at EMEC in 2009. A second 350 kW unit is for testing off Korea. preparatory to the installation of three hundred 1 MW units in the Wando Hoenggang waterway between islands off the south coast. Korea Midland Power Company aim to have the 300 MW system in operation by 2015.

A second smaller scheme with eight units is planned for E.ON to be

8.8: UK: Seagen commercial prototype with the generators raised out of the water as they would be for maintenance. (Photo courtesy of Marine Current Turbines)

installed off St David's Head, South Wales. These will be 1 MW units and are expected to be in operation by 2010.

The highest tidal range in the world is in the Bay of Fundy in Eastern Canada, at over 16 metres and it is here, and along the coasts of Maine and New Hampshire that there is growing interest in the tidal stream generator. MCT has entered an agreement with Marine Tidal Energy of Halifax, NS to develop a tidal current project for the Minas Basin at the top of the bay where the tidal current is at 3.8 m/s and some 14 billion t of water pass in or out every 6¼ hours. A Seagen Tidal unit is planned for installation there by 2011.

At the head of the Minas Basin where the Annapolis river enters from Nova Scotia there is one of only three tidal power stations operating in the world with a single 20 MW bulb turbine built into a causeway across the estuary. Completed in 1984, it generates approximatelsy 30 GWh/ year on the falling tide.

In the United Kingdom RWE have joined MCT in a plan for seven units to be installed off North Wales. This is in an area known as The Skerries at the northeast corner of Anglesey, near the port of Holyhead. It is planned to install seven units there for a total capacity of about 10.5 MW. They will be similar to the unit in Strangford Lough. Site survey is in progress with the aim to apply for consent in 2009. The plant possibly could be in operation by 2011.

Unlike other renewables tidal energy is predictable and you can catch it on the rise and fall of the tide. From that point of view it is easy to know when power will be available and when not, and it is much easier to integrate with an existing grid system than are wind generators.

Tidal current generators are of course under the sea and can therefore be considered a threat to fish populations and marine mammals such as seals, dolphins and related species. But the generators are relatively slow turning and the general turbulence of the seawater would frighten them off. Nevertheless effects on fish and the seal population of Strangford Lough will be studied during the five years of the trial.

Tidal currents would seem to be the least problematic of the renewable energy sources. But they must be designed to operate in isolation. Generators are designed for a four year maintenance cycle which is similar to that used in much of the oil and gas industry.

It is easy to think of a tidal current installation as an under water windfarm but the development of larger units is limited. The 1 MW Rotech unit is 25 metres in overall height and is designed for sites in water depths of more than 40 m. But there are the same problems of materials procurement and assembly and installation as any wind farm. Individual units weigh about 2500 t so there would be close on 750 000 tons of steel in the Korean installation, all of which has to be produced and worked into the components and at what cost in energy for this and installation on site of 300 units. At least Korea has about 30% nuclear capacity which has no emissions and can supply some of the energy to build them.

But wind power has not had a good ride. Public opinion has fought wind power because of its huge environmental impact. European Environmental policy forced the cancellation of a monster 670 MW wind farm on the Isle of Lewis, because of its environmental impact with 186 wind generators in an area of scientific interest.

Shell has pulled out of the London Array another huge offshore wind farm of 1000 MW with 341 of the 3.5 MW wind turbines, which is to be mounted 20 km off the Kent and Essex coasts across the Thames Estuary, on the grounds of costs. Centrica who already own an operating wind farm off Barrow in Furness and are building Lynn and Inner Dowsing off the Lincolnshire coast have also publically proclaimed the rising costs of assembling them and erecting them are marginalizing the economies of wind.

Since public opinion in the UK forced wind farms off shore these issues are reactions to steep increases in oil and raw material prices since 2006. But each wind generator is built up of standard items which are factory assembled under high quality control.

The wind generator is limited in size because of the length of the blades. The 5 MW machine has blades 61.3 m long, Is this the limit on size with the present technology? To go to higher output requires a taller and stronger mast with a heavier nacelle mounted on top of it with longer and stronger blades. A 7.5 MW unit to be installed at Blyth, in northeast England is 198.25 m tall to the maximum height of the blade tip. About 2% of the 2300 wind generators installed in and off shore the UK are struck by lightning every year

Assuming that the wind farm is built the completed units are not like any other generating station. A 1000 MW nuclear station can set up a contract to supply electricity at an annual availability. The operator can say when they have to shut down for maintenance and refuelling, how long it should take and at what time of year.

The operator of the 1000 MW wind farm cannot say with any certainty when it can supply its full 1000 MW and at what time of day. The climate of the country where the wind farm is sited gives some indication of when are the windiest times of year. But they cannot say with certainty that they will have an availability of 35% and when to the nearest hour they could expect to be generating. A similar proposition from a hydro station would be possible because water can be stored

So have we gone the wrong way to application? Renewables depend on natural phenomena which are variable in their occurrence and their intensity. They depend on there being other stand-by capacity to take over when they are not available which could be a combined cycle. A wind farm of less than 100 MW could be carried on the spinning reserve of the thermal plants on the system, as many of them probably are.

If, on the other hand these large wind farms had not been built but single units had been installed by specific industries then it would be a case of taking wind energy when available and drawing power from the grid at other times. A single 2 MW generator in a field which could be easily screened from some directions would be more acceptable to the public than a vast farm of fifty identical units. Wind power consumed would be free and wind power that could not be used, say at nights and weekends would earn revenue to set against the cost of power bought from the grid.

But there is another aspect of, particularly wind power, which nobody seems to talk about, not least in government where the green arguments for renewables and conservation have taken hold. Renewable energy is generated by large numbers of small units. Therefore there is a large area of land or sea which has to be occupied so that the units can be located far enough apart that one unit does not interfere with its neighbours

Secondly because they depend on natural phenomena which vary in intensity and time of operation, the design of individual units must be sufficiently strong to withstand the full range of wind speeds, which offshore would mean a Force 10 gale. A wind generator in such circumstances would not be operating but must be able to start up again as soon at the wind abated to within the range of the defined operational speeds. The blades must be strong enough and the supporting column which carries a load of several tons must not flex in the strongest winds.

Under the sea things are different. Water is denser than air and parallel units can be placed much closer together. Tidal current is measured in m/s rather than in km/h, but the units have to be securely fixed on the sea bed. The easiest way to achive this is to mount it on a heavy steel base frame whicc can just be stood on the sea bed. The actual generator installation is simpler because the generator can be designed to run at 50 or 60 Hz just like any other generating set on land. But it is mounted on a heavy, fabricated steel base.

One can certainly generate electricity from renewable energy but except for the few countries with large hydro capacity, it cannot provide total supply because of its intermittent operation. Can it be considered sustainable? Definitely not, but what has been done is done

But the problem with the new renewables is their lack of continuous availability. For a start, if the rotor diameter for a wind generator is more than 120 m, then each unit must be separated from its neighbor by at least 500 m to avoid interference between adjacent units. Thus a 180 MW offshore wind farm with five rows of ten generators would cover a sea area of 67,500 ha whereas a 180 MW combined cycle with two 60 MW gas turbines and a steam turbine would occupy about 4 ha. Furthermore the combined cycle would have annual availability of more than 90% compared with little more than 30% for the wind farm. To match the combined cycle availability of 90% would require forty-five of the fifty units of the wind farm to be running at their maximum continuous rating for 24 hours a day, and 365 days a year.

Also, in the case of the combined cycle the performance is predictable. Planners will know every day that there is gas going to the station to provide generation for the next 24 hours. But there cannot be the same certainty with wind. The weather forecast will tell them if there will be wind, but cannot be precise as to the time and the strength and duration of the wind so that they are able to deduce how many generators could operate at what time of day.

So if renewable energy has a long term future it has to be used in conjunction with the fuel based energy systems of the present. In simple terms it means using use the wind generator when the wind is blowing and solar energy during the daytime. For this reason enthusiasm for wind is greatest in countries where the governments have rejected, or are openly hostile to nuclear power: In Europe that is principally Denmark, Spain, and Germany.

The unexpected catch with renewables is the energy consumed in their construction. A 160 MW offshore wind farm would have forty-five of the current 3.7 MW wind generators. First to maufacture the supporting columns 100 m long by 10 m diameter. and forty-five sets of nacelles and blades and other equipment to be partly assembled and shipped to the port from which they will be taken for installation. Generally this is timed for final assembly offshore during the summer months when sea conditions are generally more favourable.

Compare this with a 160 MW tidal farm with 135 of the largest available units of 1.2 MW which will be mounted on 135 heavy fabricated steel bases of about 2000 t each: more than a quarter of a million tons of steel. Assuming these are built then how much energy is required to produce all this steel and fabricate it into the various columns and base assemblies. How much copper must be produced for the cables to connect the generators to a substation onshore?

Note that 160 MW is the rating of the gas-cooled Pebble-Bed modular reactor currently under development, which is housed in a reactor vessel 27 m tall by 6.2 m diameter, and would occupy a site of less than 20 hA.

If the aim of the exercise is to develop an electricity supply system which does less damage to the environment we should remember that a new technology in small units requires much more energy to be used in producing the materials and building it, than for the conventional thermal systems of the past. Damage to the environment is long term not only in the excessive use of energy to produce these small generating units but in the limited availability of the power demanding back-up from existing old fossil-fired units when they cannot operate.

All the renewable energies are electricity producers and with no potential for combined heat and power, except in the case of the biomass-fuelled steam plants. If we have to convert coal to sulphur-free diesel fuel, then the process requires heat which would normally be produced by burning some of the feedstock. Are we going to continue doing this or will we use the high temperature gas-cooled reactors when they become available to supply either steam or high-temperature air for the process, so that all of the feedstock can be used to make the product?

This is the underlying problem with renewables. The people who first advocated it and forced it upon governments are the same people who protested against nuclear energy in the past and do not see what the world is heading for with an increased population and an ever growing demand for energy.

9

The truth about America

For the last 60 years, since the end of the Second World War, European opinion has been divided between those who love America and those who hate it. This division has persisted across the years even when so many people have taken advantage of cheap air travel across the Atlantic to see for themselves.

The refusal of the United States to sign the Kyoto Treaty brought claims that they didn't believe in global warming and that they were the most pollutant nation on earth. Yet the reality of the situation is vastly different.

Environmental concern has driven American technology for more than forty years but the fact is that the United States is one of a few very large countries with widely dispersed population centres, which tends to hide the results. It's great industrial strength was developed on the east and west coasts during the Second World War which for them was fought overseas in the western Pacific and Europe.

The industry of the United States has been one of mainly innovation. The steam turbine, and gas turbine were developed in Europe, but it was American companies that came to dominate the global market for these products after 1950. But the environmental aspects of energy production and use has been very much an American achievement.

Unleaded gasoline and the exhaust catalyst first appeared for new cars in the United States in the early 1970's and the first flue gas desulphurization units appeared on new coal-fired power stations before the end of the decade.

Combined cycle plants in the United States have catalysts in the heat recovery boilers to reduce NOx emissions from 25 vppm or less at the gas turbine exhaust flange to 3 vppm at the top of the stack. This has had a significant impact in the reduction of emissions of nitrogen oxides

A further achievement is that in New York, one of the most populous states, greenhouse gas emissions in the electricity supply sector in 2004

TABLE 9.1: US GENERATING CAPACITY BY FUEL

Fuel	Units	Rating MW	Comments
Biomass	25	302.7	Bagasse, rice husks, etc
Other biomass	95	344.3	
Blast furnace gas	34	1005.5	Steel industry
Black liquor	164	4108.2	Pulp and paper industry
Geothermal	217	3170.9	
Landfill gas	688	1063.5	Anaerobic digestion
Lignite	30	14788.2	Texas and North Dakota
Municipal solid waste	96	2671.3	Refuse incinerators
Natural gas	5460	426751.1	
Industrial waste gases	71	1557.4	Hydrogen, propane, etc
Nuclear	104	105584.9	
Purchased steam	23	476.4	
Diesel fuel oil	3447	28524.4	Mainly small engines
Distillates	84	2676.1	Gas turbine peakers
Residual oil	178	31271.8	
Waste oil	6	79.5	
Petroleum coke	32	1766.4	
Bitumenous coal	948	181408.7	
Sub-bitumenous coal	473	132065.2	
Other coals	41	8533.8	Colliery wastes, briquettes
Solar	18	411.4	
Scrap tyres	2	54.8	
Water	4138	1648768.4	Includes pumped storage
Wood	174	2984.5	
Waste wood	13	359.2	
Wind	9661	11334.4	Total units on 343 sites
Total	26222	2612063.0	

were significantly below the level in 1990 on all counts. This is contrary to the national trend, although across the whole country emissions of sulphur and nitrogen oxides have markedly decreased in the past fifteen years. This could be due to the greater share of power generation being taken by gas-fired combined cycles and the improved performance of the nuclear power plants.

Besides this there is a considerable diversity of fuels in use throughout the power generation industry, which must include the large number of industrial producers, of which steel, and pulp and paper are significant in producing much of their own fuel. Twelve steelworks in five states generate 1000 MW with blast furnace gas, all to power steam turbines.

Similarly in a pulp and paper mill, once all the solid fibres have been removed to make the paper, you are left with black liquor which can be burned to produce steam for stock preparation and drying of the paper. Other plant debris such as bark and twigs can also be burned to produce more steam. In total 68 mills generate 4102 MW using mainly black liquor as the fuel.

At the end of 2007 there were 26 222 generating sets available to operate with a combined nameplate rating of 2 612 063 MW. The fuels are indicated in the table and it can be seen that there is a large choice of waste fuels and particularly a significant output of energy from refuse disposal.

Surprisingly for a country so dependent on road transport there are only two plants using scrap tyres as fuel. One is owned by CMS Energy at Exeter CT, with a 31.5 MW steam turbine set serving the New England Market: the other at Ford Heights, IL, 25 km south of Chicago has been shut since 2004 when the original owners filed for bankruptcy. Geneva Energy took over the plant and was facing public protest against its plan to restart in 2008 to burn the equivalent of 700 tyres per hour to produce 20 MW for sale into the public grid.

But it is these small contributions to electricity supply which are the most interesting because they show a determination to capture the energy from waste materials.

As would be expected the major population centres are in the forefront with California and the east coast states, also Pennsylvania and Illinois. The oldest refuse incinerators date from around 1950 at the Elk River, Red Wing and Ilmarth sites in Minnesota. But the rest date from after 1980 suggesting that the PURPA legislation might have been a stimulus to their construction.

Landfill gas has been exploited for about 25 years with the first projects going into southern California in 1984. The gas is produced by anaerobic digestion of animal and vegetable wastes deposited in the landfills.

Landfills designed to exploit digester gas construct clay cells into which the refuse can be placed so that the gas can be concentrated and extracted through one or more pipes. Because of the low gas pressure the majority of sites use reciprocating gas engines ranging in size from 500 to 1500 kW capacity.

Once full, each cell is capped and when gas production ends it is shut off and the next full cell is connected into the network. Then, when there is no more space to expand the landfill and all gas production has ceased, the pipes and other equipment can be removed and the land returned to

TABLE 9.2: ENERGY FROM MUNICIPAL WASTES

State	Landfill Gas Sites	MW	Incinerators Sites	MW	Total MW
Arizona	1	5.00			5.00
California	36	212.90	3	71.60	284.50
Connecticut	2	5.70	6	215.80	221.50
Florida	4	18.00	10	499.80	517.80
Georgia	1	2.40	1	5.50	7.90
Hawaii	11	63.70			63.70
Iowa	1	6.40			6.40
Illinois	26	107.40			107.40
Indiana	4	14.60	1	6.50	21.10
Kentucky	3	9.60			9.60
Maine	5	65.60			65.60
Massachusetts	9	31.40	5	295.50	326.90
Maryland	3	9.50	3	133.50	143.00
Michigan	14	80.40	5	90.10	170.50
Minnesota	1	48.90	9	134.90	143.80
North Carolina	2	12.70	2	10.50	23.20
Nebraska	1	3.20			3.20
New Hampshire	3	13.60	2	18.50	32.10
New Jersey	7	25.20	5	177.30	202.50
New York	11	76.40	10	327.30	403.70
Ohio	1	3.80			3.80
Oklahoma	1	16.80			16.80
Oregon	2	2.40	1	13.10	15.50
Pennsylvania	10	131.60	3	247.60	379.20
Rhode Island	1	7.60			7.60
South Carolina	2	8.80	1	13.00	21.80
Tennesee	2	5.00			5.00
Texas	8	37.80			37.80
Utah	1	1.60			1.60
Virginia	5	14.40	2	213.00	227.40
Washington	2	13.20	1	26.00	39.20
Wisconsin	12	52.40			52.40
Total		963.00		2647.20	3610.20

agriculture, or made into a golf course, or put to some other use.

Of course refuse incineration and landfill schemes are found in other countries but these began as municipally owned schemes in parts of Europe where the electricity supply system was mainly in state ownership. Denmark, for example, is one of several countries in northern Europe which for many years has had refuse incinerators in the major cities which are not only generating electricity but feeding heat to the local district heating system.

In the United States, however, district heating is on a small scale and mainly built up from earlier steam systems based on boiler plant alone.

Combined heat and power is largely confined to industrial power schemes and the few district heating schemes are mainly downtown networks in some of the northern state capitals dating from the end of the 19th century. The big spur to combined heat and power was the Public Utilities Regulatory Powers Act of 1979. This laid down the conditions for industry to generate its own heat and power and sell the surplus power to the public utility at market rates.

NRG Energy LLC, based in Princeton, NJ, have through their daughter company, NRG Thermal, bought into six schemes including one combined heat and power. scheme at Dover, Delaware. It dates from 1984 when General Foods built a 16 MW coal-fired plant to supply power and process steam to their factory. When Kraft Foods bought them out in 1996 they sold the power plant to StatOil, of Norway, who in 2000 sold it to NRG Thermal. Two GE LM6000 gas turbine sets were then added to provide peaking power to the city of Dover, with the existing plant supplying power and up to 31.8 t/h of steam to Kraft.

Harrisburg, PA, has had a steam system through the downtown area since 1887, which was taken over by Pennsylvania Power and Light (PPL) in 1929. There are now three plants serving the system: a large steam plant with four dual fuelled gas/oil boilers, three of which produce 61.5 t/h with the fourth supplying 18 t/h. In 1986 the second plant was built with two 6 MW diesel generator sets with heat recovery boilers supplying a further 3.2 t/h. Since the beginning of 2006 the city's refuse incinerator has supplied 68 t/h to the system. The third plant supplies chilled water for air conditioning from two 1200 t electric chillers.

Effectively NRG run parallel systems for heating and air conditioning which they supply to an area of about 2.6 km^2. Among the buildings served are the State Capitol complex, the Harrisburg Hospital, the Harrisburg University of Science and Technology, an hotel and a shopping mall, also the Taylor Wharton factory which manufactures heavy steel cylinders for the transport of acetylene and other industrial gases. In fact Harrisburg is typical of these district energy schemes which supply a few large energy users within a compact area.

The disadvantage is that these schemes use steam whereas the much more extensive European schemes mainly use pressurised hot water. One scheme which has introduced hot water alongside chilled water and steam supplies is a district heating and cooling system in Minneapolis.

The scheme there is not combined heat and power, but it provides district heating and cooling from the Downtown plant to some 100 buildings with 4 million m² of floor space for heating and 2 million m² of space in 40 buildings receive district cooling. The Fairview Augsburg plant provides steam service to the campus of Augsburg College and the University of Minnesota Medical Center.

So there is some district heating in the United States, but largely steam-based and therefore concentrated among groups of large energy users on small networks in city centres, and since these have the largest air conditioning loads in summer, there is scope for some to use steam to drive absoption chillers on the premises, while others take chilled water from a central plant.

Far more significant is the fact that industrial combined heat and power schemes provided 4.25% of the total public electricity supply in 2008. This would be surplus power sold by the operators, some of which have quite big generating plants. up to 250 MW.

Agriculture is another interesting group of producers. With a population of 300 million the United States is a large producer of sugar, mainly in Florida, Louisiana and Texas, and of rice in California and Louisiana. In these States small steam plants are fuelled with bagasse, rice husks, straw and wood chips. Ten power plants produced 282.7 MW and those run by the sugar producers also supply steam to the sugar refineries.

One of the most interesting is the Chateau Energy 21 MW plant in California's Imperial valley for which the fuel is cattle manure. It is so far the only one of its kind. Across the United States cattle are transported to feed lots near the major cities where they are finished off before slaughter. These sites produce a lot of manure and the Mesquite Lake Resource Recovery plant about 150 km east of Los Angeles was designed to burn 600 t/year of manure, and would be the first of several such plants to get rid of a mounting problem of disposal across the country.

A newcomer to biomass fuel is the chocolate industry. Branch companies of European producers have have imported chocolate prepared from the cocoa beans from which they made the various products for their local markets. Now Sprüngli Lindt in Portsmouth, NH have decided to make their own chocolate from imported beans and have entered an agreement with Public Service of New Hampshire. Lindt would supply the discarded shells which PSNH would co-fire with coal at their 145 MW Schiller power station. The utility were in early 2009 performing test firings of the new fuel mixture.

TABLE 9.3: SOME BIOMASS POWER PLANTS

Operator	Location	Fuel	Output MW
Agrilectric Power Partners	Calcasieu, LA	Rice husks	13.6
Archer Daniels Midland Co	Enderlin, ND	Wood chips	9.8
Chateau Energy Inc	Mesquite, CA	Manure	21.0
Gay & Robinson Inc	Kauai, HI	Bagasse	4.0
Hawaiian Sugar Company	Maui, HI	Bagasse	36.1
M A Patout & Sons Ltd	Jeanerette, LA	Bagasse	3.0
New Hope Power Partners	Okeelanta, FL	Bagasse	74.9
PSNH[1]	Portsmouth NH	Cocoa shells	145.0
Rio Grande Valley Sugar	Santa Rosa, TX	Bagasse	7.5
United States Sugar Corp.	Clewiston, FL	Bagasse	45.7
United States Sugar Corp.	Bryant, FL	Bagasse	28.5
Wadham Energy Partners	Williams, CA	Rice husks	28.6
Total			282.7

[1]Co-firing cocoa shells with coal.

In total, the net electricity production in the United States at the end of 2006 was 4065 TWh, an increase of 0.25% over the previous year, and of 18% over the previous ten years. So the country is very much in tune with the developed industrial nations of Europe which have shown similar low growth in electricity demand over the same period. However per capita electricity consumption is 11000 kWh/year which is not the highest in the world. In Canada per capita consumption is 14000 kWh/year while in most European countries consumption is less than half the US figure.

The largest electricity production in the United States was from coal, followed by natural gas, with nuclear running a close third and hydro fourth. All other sources, including oil and wind each contribute less than 2.5% of the total. But nuclear, hydro and wind together contribute 28.85% entirely free of emissions, and this total is bound to rise. Several of the New Start licenses have been issued for nuclear plant, while existing nuclear stations are gaining power through turbine upgrades and improved fuel burn-up rates.

There are 104 operating nuclear plants which provide almost 20% of the national electricity supply and many have in the past eight years been given life extensions which will see most of them still operating beyond 2030. Tennessee Valley Authority are building the second of two reactors at their Watts Bar site. The 1200 MW PWR is expected to go into operation in 2012. The first of the new reactors, which is expected

9.1: Grand Coulee Dam, USA. Three plants generate 6500 MW with a pumping station to irrigate central Washington State. (Photo courtesy of Bureau of Reclamation)

to be in Levy County, FL, will probably be complete and in service in 2017.

The other aspect of American energy is hydro power which was the main source of electricity over much of the country in the first half of the 20th century. The Hoover dam on the lower Colorado River completed in 1938 was the catalyst for the development of the desert states in the Southwest.

The level of Lake Mead however has dropped with the combined effects of drought and increased water demand in Las Vegas over the last eight years, and unless there is more rain upstream it may eventually threaten the operation of the power station.

In 1942 the even larger Grand Coulee dam was completed in Washington State on the Columbia River, and although intended for both power generation and irrigation, the war effort put the emphasis on power generation and irrigation services were not started until after the end of the war. Later a third power plant was added and the pumping station was doubled using reversible pump turbines in the new section so that it can also function as a peaking plant.

Sixty-five years later there are three power plants at Grand Coulee with twenty-four generators at a total rating of 6500 MW and twelve pumping sets which provide irrigation to 245 000 ha of farmland in central Washington State through a system of reservoirs and canals.

9.2: Hoover Dam, USA. Completed in 1938 near Las Vegas, NV. Long term operation of the power plant is threatened by low rainfall upstream in recent years, and high water demand in the region.

At the end of 2006 there were in operation in the United States 4142 water turbines with a total capacity of 96 923.7 MW which were reinforced by hydro power exported from Quebec. Manitoba, and British Columbia, the three provinces of Canada with large hydro power reserves. Existing US hydro stations the oldest of which date back to the 1920's are also having turbine refurbishment and generator upgrades.

So to argue as Green activists do, that the United States does not believe in Global Warming and is polluting the atmosphere for all its worth, is blatantly untrue. There are no countries without a large component of hydro or nuclear plant that can point to 30% of their electricity supply being totally emission free.

However the Bush Administration set in train a reduction of electricity consumption by legislating for the change over from incandescent to low-energy fluorescent lighting. It has also licensed new designs of nuclear reactor and reworked the licensing procedure into a single format which must be applied by the licensee within 20 years.

Another legislative measure is the extension of the federal clean energy tax credit scheme. The Production Tax Credit which pays $20/MWh to producers of renewable energy, including wind, biomass, geothermal, and hydro power, continues for one more year and for each case runs for ten years from the date of start-up. The investment tax credit for solar and fuel cell technologies is extended for eight years.

TABLE 9.4 GROWTH OF ELECTRICITY SUPPLY (TWh)

Year	Coal	Oil	Gas	Nuclear	Renewables Hydro	Other	Total
1955	301.4	37.1	95.3		116.2	0.3	550.3
1960	403.1	48.0	158.0	0.5	149.4	0.2	759.2
1965	570.9	64 .8	221.6	3.7	197.0	0.4	1058.4
1970	704.4	184.2	372.9	21.8	251.0	0.8	1535.1
1975	852.8	289 .1	299.8	172.5	303.2	3 .4	1920.8
1980	1161.6	246 .0	346.2	251.1	279.2	5 .5	2289.6
1985	1402.1	100.2	291.9	387.3	284.3	10.7	2476.5
1990	1594.0	126 .6	383.2	573.4	292.9	67.9	3038.0
1995	1709.4	74 .6	510.0	670.7	310.8	78.1	3353.5
2000	1966.3	111.2	615.0	751.4	275.6	85.7	3802.1
2005	2013.2	122 .5	774.3	775.4	270.3	99.7	4055.4

Immigration, industrial growth and rising living standards have brought about an eight-fold increase in electricity demand in the past 50 years. In the 1950's coal and hydro power were the main sources of electricity with oil and gas concentrated mainly in the south and west where there were oil fields but as yet no coal fields developed. The first gas turbines were being used to boost the efficiency of oil fired plants by using exhaust gas to heat feed water. In 1957 the first nuclear plant a 68 MW PWR, by Westinghouse went into service at Shippingport, PA.

At that time the electricity supply industry was organised by State with municipal and investor-owned utility groups. But there were two government organizations who also generated and transmitted energy and which had developed out of the Roosevelt New Deal in the 1930's. The Tennessee Valley Authority headquartered in Knoxville, TN was set up as a utility company to build a series of hydro stations on the Tennessee River to supply consumers in Tennessee and neighboring States in the Southeast.

The Bonneville Power Administration, headquartered in Portland, OR, was set up as a Marketing Agency for Federally owned power plants, mainly in the western states, and operated by the US Army Corps of Engineers and the Bureau of Reclamation. Interconnection was not across the whole country but within three large regional groups Eastern and Western separated by the Missisippi River, and the state of Texas.

Forty years ago acid rain led to the discussion of burning western low-sulphur coals in eastern power stations. FGD, with a commercial product, gypsum, at the end of it, was a better solution than a procession of 2 km long freight trains carrying the coal eastwards and empty wagons the other way which would dominate freight traffic on the transcontinental routes. Furthermore the sale of the gypsum byproduct would provide income to help pay for the FGD installation.

Coal-fired plants were built close to the western coal fields to supply the western states and as these were completed so the share of oil in electricity supply started to decline. But since then, environmental pressures have mounted against coal, which in 2006 provided 49% of the electricity generated in the United States. By 2030 demand is expected to have grown to 5800 TWh with coal still accounting for the largest share. But what sort of plant will provide this energy, will it be aging subcritical steam plant without carbon capture or new IGCC (Integrated Gasifier and Combined Cycle) schemes?

Some 68% of the available coal-fired capacity of all types at the end of 2005 was at least 35 years old and much of it considerably older. In fact nearly all of this old plant is operating on low-pressure sub-critical steam cycles with efficiency at best of 35%. So a particular issue which must be addressed with greater urgency as time passes, is the replacement of this capacity, and with what.

IGCC as presently conceived is less efficient than coal-fired supercritical steam plant, even less so if equipped with carbon capture. IGCC needs further development beyond even the use of the bigger H-class gas turbines, and if CCS is included then the gas turbines must be able to burn hydrogen at the higher temperatures with the same emissions performance as the current F-class gas turbines.

In 1962 a study by Bechtel for a new power plant in Texas found that to build a 120 MW steam power plant, but with a gas turbine providing the feedwater heating would cost $3.60 less per kW installed than a 120 MW station with a conventional reheat steam cycle. No such cost advantage has come with the environmental additions to the modern steam plants of ten times the size. The addition of carbon capture will add further to the cost.

There has for a long time been seen a need to reduce fuel imports. Most natural gas has come from Alaska and Canada, but an increasing volume of liquefied natural gas is coming in from Trinidad, Norway, Indonesia, the Middle East and Nigeria.

The planned surge in gas-fired capacity in 2001 was brought to a halt by rising prices and the bankruptcy of Enron a major gas supplier and project developer. As the Industry deregulated a number of generators invested in generating companies overseas to gain experience of the sytem in other countries, mainly the UK, and Australia. Similarly companies from deregulated Europe moved into the United States,

among them National Grid and Centrica from the UK, Suez Tractebel
from Belgium, and E.ON from Germany. Some bought into utility
systems and others bought specific power plants and have built more.
International Power in the UK set up American National Power (ANP)
headquartered in Houston, TX.

The European maufacturers also bought into American industry. The
most notable example was Siemens purchase of the Westinghouse non-
nuclear business in 1998 and also the Turbocare gas turbine maintenance
company. Turbocare is now a truly global maintenance group and not
just covering only Siemens gas turbines. Allison was taken over by
Rolls-Royce. Westinghouse nuclear went to British Nuclear Fuels but in
2005 was sold to Toshiba, since when they have got their first order for
the Advanced PWR in China.

ABB before they were tied into Alstom had taken over Combustion
Engineering boiler division in Windsor, CT, and had built on the CE
experience with once through boilers to develop the unit for the GT24
single shaft combined cycle blocks. There are some 28 of these plants in
operation in the United States and Mexico, all of a similar design with a
centrally-mounted generator and a 2-pressure Benson boiler with a high
pressure output at 160 bars, 560°C driving a high-efficiency 2-speed
steam turbine.

The first was installed at Agawam, MA, in 2001, for Berkshire
Power. The 14 ANP sets are in Texas at Midlothian (6) and Hay (4); and
in Massachusets at Blackstone and Bellingham, with two each. The only
variations are in the cooling arrangements and some insulation to meet
noise criteria at the site boundary.

The 104 currently operating nuclear reactors have no emissions
and have shown considerable improvement in performance in recent
years and stability of generating costs compared with the fossil-fuelled
alternatives.

One more of the first series of reactors has to be completed as
Watts Bar unit 2. Although it first appeared in TVA plans back in 1970,
construction of the two reactors was delayed not so much because of
anti-nuclear protest, as for the more mundane situation of low growth in
demand for electricity. This has not deterred the nuclear industry from
developing new reactor designs. These are all being produced in one or
more standard designs for the global market, and to date four designs
have received certification.

GE produced an upgraded design of their reactor as the Advanced
Boiling Water Reactor and installed the first two in Japan at Tokyo
Electric Power's Kashiwazaki Kariwa station as units 6 and 7 each at

1315 MW, completed in November 1996 and July 1997. A third Japanese unit was later sold to Chubu Electric rated at 1325 MW and has been in service since January 2005. For the US and other markets the BWR has been further developed as the Economically Simplified, ESBWR.

Westinghouse has developed 600 and 1100 MW versions of its Advanced PWR and it is the larger unit which has achieved the first sales for two reactors on each of two sites in China. Fourteen units are covered by New Start Licenses and the reactor design has also passed the initial evaluation of the British Nuclear Regulators, for possible use in that country's new nuclear power programme.

Some of the new licenses in the United States have been issued to operators of existing plants to install further units on their sites, with TVA and Entergy earmarked for the first commercial examples of the AP 1000 and ESBWR designs, respectively.

Twenty years ago people were running around saying that no more nuclear power plants would ever be built in the United States. A combination of Green protest and low rate of growth in electricity demand meant that construction of those power plants that had survived the various licensing hurdles slowed right down which simply encouraged people to think that way. The industry was kept alive with a few orders in the Far East, and one in the UK, together with service and uprating of the operating plants.

In June 2005, President George W Bush made an official visit to the nuclear plant at Calvert Cliffs, MD. Some 26 years earlier the Carter adminstration had called the nuclear power programme into question with the cancellation of the Barnwell reprocessing plant and forcing a once-through fuel cycle on the operators.

This was all about to change as in a speech at the power station the President announced his determination to revive the nuclear industry so that it could play its full part in helping to clean up energy production and reduce fuel imports.

Two months later, in August 2005 the Energy Policy Act was passed. This encourages investment in non carbon dioxide emitting technologies. In particular it provided incentives for the construction of new reactors in the form of federal tax credits and loan guarantees for the first six reactors to be built, together with insurance against regulatory risks that would cover expenses up to $500 million caused by any licensing delays for a new project.

The licensing procedures have also been simplified. Instead of separate licenses for construction and operation, the single new-start license has been created which is a single license to construct and operate a plant on the selected site. On completion of construction there will be supervised tests leading to commercial operation. The whole process should see the first of the new nuclear plants in service by the end of 2017.

At the end of 2008, applications for Construction and Operation Licences had been filed with the Nuclear Regulatory Commission at seventeen sites. Of the twenty-six reactors four are the 1600 MW Areva EPR, of which Calvert Cliffs, MD is the lead unit, and two are the 1700 MW Mitsubishi USAPWR. But what this shows is that the nuclear business is becoming more international with US versions of the EPR reactor and Mitsubishi's Advanced PWR which are expected to receive NRC certification by the end of 2009.

The problem with the American Industry at the beginning was the general reluctance to accept turnkey contracts. The utilities had always ordered boilers and steam turbogenerator sets under separate contracts so why not a reactor, after all it was only a special type of boiler. Two of the original reactor vendors were in any case boiler companies: Babcock & Wilcox, and Combustion Engineering both offered alternative PWR designs, in addition to fabricating reactor vessels and steam generators for GE and Westinghouse. So there was no one standard design except at sites such as Duke Energy's Oconee, NC, and Pinnacle West's Palo Verde, AZ, plant each with three identical generating units .

With the lack of recent construction and the improved performance of the existing reactors, there has been a significant change in public opinion. There is now a clear majority of two thirds to one in favour of continued nuclear development.

For a country importing increasing quantities of fuel, nuclear power has two specific advantages. First is that uranium is an indigenous resource and that, together with the retired nuclear weapons material there is a huge potential for nuclear power, and particularly with mixed oxide fuel, but only to generate electricity. Second there are no emissions from operation. So that if the country as a whole were to achieve 60% nuclear power supply like some of their Asian allies and competitors aim to do, this would make a significant change in air quality.

But however big a majority in favour there is still 30% who are against nuclear power and are still active. They can create more protest and this could be why the governement has streamlined the licence procedure into a single Construction and Operation Licence. and set up an insurance scheme to compensate for licensing delays for the first of the new reactors.

The greatly improved performance of the American nuclear fleet in

TABLE 9.5: COL APPLICATIONS AT END OF 2008

Owner	Site	No.	Type
STP Nuclear Operating Co.	Matagorda County, TX	2	ABWR
Tennessee Valley Authority	Bellefonte, AL	2	AP1000
Dominion Energy	North Anna, VA	1	ESBWR
Duke Energy	Cherokee County, SC	2	AP1000
Progress Energy	Shearon Harris, NC	2	AP1000
Entergy	Grand Gulf, MS	1	ESBWR[1]
UniStar	Calvert Cliffs, MD	1	EPR
Southern Nuclear	Vogtle, GA	2	AP1000
SCE&G V.C.	Summer, SC	2	AP1000
Progress Energy	Levy County, FL `	2	AP1000
Exelon Generation	Victoria County, TX	2	ESBWR
Detroit Edison	Fermi, MI	1	ESBWR
Luminant (TXU)	Comanche Peak, TX	2	USAPWR
Entergy	River Bend, LA	1	ESBWR*
AmerenUE (UniStar)	Callaway MO	1	EPR
UniStar (Constellation)	Nine Mile Point NY	1	EPR
PPL (UniStar)	Bell Bend PA	1	EPR

[1] COL Application with ESBWR but delayed to consider other reactor types.

the last fifteen years has been attributed to the deregulation of Electricity Supply which among other things concentrated the nuclear operators into specialist groups which were better able to pool maintenance services and spare parts and so improve availability, which now averages over 90%.

During this time the nuclear industry has concentrated on the operation and inprovement of the existing reactor fleet. But with the increase of nuclear construction around the world Areva, going back to their origins in Germany and France, have between them in total built 102 reactors of which all but ten are in Europe, with the majority in France

Areva have an expanding industrial base in the United States comcentrated in the fuel cycle. In partnership with the Shaw Group they are building a Mixed Oxide fuel plant at the Savannah River site in South Carolina. Construction started in 2007 and production is expected to start in 2016. The origin of the plant was a US agreement with Russia whereby each country would decommission 34 tons of military-grade plutonium and use it to make mixed oside fuel for power reactors. Areva have already made sample mixed oxide bundles at their Cadarache facility in France in 2004 and 2005 which have been supplied to Duke Energy's Catawba power plant in North Carolina.

9.3: Dresden, IL: One of six nuclear plants in the State has two GE 850 MW BWR with operating license extended to 60 years. (Photo courtesy of Exelon Corporation)

In May 2008 Areva chose a site in Bonneville County, ID, some 30 km from Idaho Falls for a centrifuge enrichment plant. These are the first steps following the setting up of the Global Nuclear Energy Partnership (GNEP) in 2003 to repair the damage done to the industry in 1979 under the Carter presidency.

GNEP is the start of a global fuel cycle industry to reprocess spent fuel for all operators and make mixed oxide fuel to extend the use of the uranium resource, and also to consume the material of the degraded warheads of the nuclear weapons states.

But with the licensing of new reactor designs which are already selling abroad, the range of designs is increasing into smaller units with a wider possibility for application. The most important of these is the Pebble-Bed Modular Reactor, which is under development in South Africa and China.

The reactor is being studied both as an electricity generator and as a heat source for coal gasification and oil shale processing. It is the subject of an NRC license application which is expected to be completed in 2012 when the first unit should be nearing operation in South Africa, after which an American prototype will be built at Hanford, WA.

The importance of this reactor is its low capacity and high operating temperature which lends itself to coal gasification and other similar processes which at present derive energy from burning some of the feedstock. So the importance of an intrinsically safe reactor for industrial processes cannot be ignored.

9.4: Diablo Canyon, CA with two Westinghouse 1125 MW PWR completed in 1985. Station is deigned to resist 7.5 Richter earthquake.
(Photo courtesy P G & E)

IRIS, a compact 300 MW PWR design is being developed as a
GNEP project under the leadership of Westinghouse with contributions
from Italy, the UK, Croatia Lithuania and Russia. Work at present
includes component design and testing, and various design studies of
applications. Design certification by NRC is expected in 2012. The
Babcock Modular reactor, a 125 MW PWR is expected to follow at
about the same time.

The Next Generation Reactor is an American High Temperature Gas-
cooled reactor which is aimed at a wider spread of applications, such
as coal gasification and oil sands processing. It may still be ten years
into the future, but the logic of this development is to bring nuclear
combined heat and power to industry.

But a wider application of nuclear energy raises other issues. It had
taken thirty years to produce new reactor designs based on the original
PWR and BWR systems, and for much of that time there was general
hostility to nuclear power. It is only since Kyoto that TVA has completed
Watts Bar 1, and the second reactor is due in service in 2012. But
deregulation and the concentration of some nuclear projects into larger
companies has seen improved performance and low generating cost
which have revived interest with the need to combat climate change.

The Bush Administration which left office at the end of 2008 is largely
responsible for the improvement of licensing terms which have resulted
in the applications for seventeen projects with twenty-six reactors. In so
far as Energy is an election issue there is a general concensus that oil

and gas imports should not be increased and if anything reduced. So there is a considerable amount of research and development under way to produce cheaper and cleaner energy with fewer emissions.

Besides the nuclear activities, there are projects to develop gas turbines with higher efficiency and lower emissions when burning hydrogen, but there is surely scope for these designs to be carried over to the natural gas-fired combined cycle as well. We have yet to see 60% achieved with the current H-class gas turbines.

There are also the trial operations for carbon capture and storage the first of which started a twelve month trial at We Energy's Pleasant Prairie station, Wisconsin, in February 2008. It is unlikely that any plans for a full-scale trial on a power station will be announced until the results of these trials are known.

But for the present the total forecast for future energy demand sees a 16.7% increase up to 2030. Use of natural gas will rise up to 2020 but then growth in demand will reduce to become only 3% higher in 2030, and with imports of natural gas reducing by 8% as compared with 2006.

Nuclear energy is forecast to rise by 16.6% with hydro power increasing by 3.8%. But already plans for 26 reactors have been lodged with the Nuclear Regulatory Commission. These are all large units for electricity generation. As can be seen in the table these range in size between 1100 and 1700 MW. The majority are PWR with sixteen AP 1000 on eight sites and four of the Areva EPR design on four sites.

One particular project at Grand View, ID is planned as a combined heat and power scheme with steam being supplied to produce ethanol. Thirty years ago a 1200 MW nuclear plant at Midland, MI, with two PWR was planned as a combined heat and power scheme to supply steam to the neighboring Dow Chemicals plant; the project was abandoned when almost complete and it was later converted to a combined cycle by coupling twelve gas turbines and heat recovery boilers to the modified steam turbines.

Grand View will be the first of a kind as a provider of emission free process heat and remove the need to burn some of the feedstock to provide the process energy. This will be more significant with the arrival of the smaller high temperature gas-cooled reactors which are being designed both for electricity generation and as a combined heat and power system for processes such as coal gasification.

Biomass is to increase by 176.2%, and other renewables, principally wind and solar 178.4%, but starting from a relatively small base. It is not clear whether the biomass is farm and forestry and agricultural

TABLE 9.6: GENERATING PLANT FORECASTS (GW)

	2008	2010	2015	2020	2025	2030
Coal	306.7	311.4	319.3	338.5	367.6	401.5
Oil and Gas	118.0	118.0	93.2	93.0	92.6	92.6
Combined Cycle	154.1	158.2	159.9	164.2	173.3	177.5
GT and Diesel	133.5	134.5	127.1	129.2	140.9	161.8
Nuclear Power	100.5	100.9	102.1	110.9	115.7	114.9
Pumped Storage	21.5	21.5	21.5	21.5	21.5	21.5
Renewables	104.9	110.9	116.6	122.9	127.5	131.8
Distributed Power	0.0	0.3	0.9	2.7	5.9	9.8
Industrial CHP	40.3	40.3	41.0	41.0	41.0	41.0
Subtotal	939.1	955.7	940.6	982.8	1045.0	1111.4
Planned Additions	26.8	44.0	63.7	105.7	137.9	172.0
Planned Closures	2.7	3.6	38.9	39.5	40.0	44.8
Total Capacity	979.4	996.0	981.6	1023.8	1086.0	1152.4

wastes, including bagasse and black liquor, or if it includes fuel crops to make bio-diesel. But if we look at the total domestic resources of nuclear, hydro, biomass, and other renewables, which currently account for 29.1% of the total electricity production, their share stays roughly the same in 2030..

Electricity production from all sources was 3975 TWh in 2008 and forecast to reach 4777 TWh by 2030 at an annual rate of no more than 1.1% per year. In 2008, coal supplied 50% followed by nuclear (20.1) followed by natural gas (17.25). In 2020 coal will have declined to 46% with nuclear (19.8) and natural gas (15.8) Renewables, predominantly hydro, will have risen from 8.9% of supply in 2008 to 12.5% in 2020. Ten years later coal will account for 48.6% with nuclear (18.9) natural gas (16.4) and renewables (12.7)

The lower nuclear output is surprising given that at the end of 2008 applications for seventeen projects had been received with a total name plate rating of 31.8 GW with another 2.7 GW to be obtained by turbine and fuel upgrades at existing plants. The oldest and smallest plants totaling about 4.5 GW will be shut down and decommissioned. The nuclear total could increase if all the planned projects are built before 2030 and a lot more coal-fired plant is shut down.

Railways are predominantly diesel-powered with only the main high-speed line from Boston through New York to Washington DC electrified. Several States have discussed high-speed rail links but have met with problems of funding, and environmental objections. A high speed service between Houston, Dallas, Fort Worth, and Austin was killed off by the local airline Southern Air who served all four cities. Perhaps with higher costs of oil affecting airline economics and the costs of motoring, there may in future be the possibility to build some of these high speed rail links in the more populated areas of the country.

The problems of coal are compounded by the fact that the utility companies have been slow to adopt the supercritical steam cycle though some recent orders have been placed. The real problem is with IGCC and there is increasing public comment against it because even without Carbon Capture and Storage, the performance of the few operating plants is significantly below that of current supercritical steam plants.

Feasibility studies have been published but no IGCC has yet been ordered for a 630 MW plant based on F-class gas turbines. since the demonstration projects were constructed more than a decade ago. A utility faced with the choice of an ultrasupercritical steam plant with an efficiency of 46% or an IGCC at 38% is surely going to opt for the steam plant because it will burn 20% less coal which means 20% lower emissions. and have a smaller supporting energy load for coal delivery and disposal of ash and is an altogether simpler and easier plant to maintain.

The existing coal-based IGCC plants have shown that they can handle a variety of coals either separately or mixed with petroleum coke. But the next problem is to integrate the plant with a carbon capture system. This is easier said than done and at the end of 2008 there is still no carbon capture system in operation or planned that can handle the entire flue gas output of a coal-fired power station.

In January 2008 the Department of Energy withdrew funding from Futuregen, an ambitious international project to build and test a zero-emissions coal-fired power station at Mattoon, IL. The argument at the time was that there was too much concentrated on this one project when Carbon Capture and Storage (CCS) was to be studied for all current coal-fired plants.

Four months later revised plans were drawn up to accelerate the near-term deployment of advanced clean coal technology by equipping new IGCC or other clean coal commercial power plants with CCS technology. Proposals were invited for plants which would be able to sequester at least 1 million t/year of carbon dioxide. At least 50% of the output of the power station must be used to produce electricity; and the gross electric output must be at least 300 MW.

TABLE 9.7: ELECTRIC OUTPUT FORECASTS (TWh)

	2008	2010	2015	2020	2025	2030
Coal	1968	2006	2065	2094	2116	2320
Petroleum	42	42	44	44	45	46
Natural gas	686	635	616	685	826	785
Nuclear	799	809	831	862	867	905
Pumped storage	1	1	1	1	1	1
Renewables	356	410	472	542	581	607
Total	3852	3904	4029	4228	4436	4665
Industrial CHP	156	143	148	149	150	146
Total generation	4008	4047	4177	4377	4586	4811
Less plant auxiliaries	34	34	33	34	34	33
Net output to Grid	3975	4013	4143	4344	4553	4777

In addition, the projects must be designed to aim for 90% capture of carbon content in the syngas or flue gas but must achieve a minimum capture rate of 81%. The plant must remove at least 90% of the mercury emissions based on the content of the coal, and 99% of the sulphur, and reduce nitrogen oxide and particulate emissions to very low levels.

Where carbon separation is applied in the chemical industry the volumes of gas are relatively small. However the Great Plains Synthetic Fuel plant at Beulah, ND separates out 6000 t/day of carbon dioxide, approximately 2.2 million t/year, which since 2000 has been piped 100 km north to Weyburn, Saskatchewan where it is used for enhanced oil recovery.

If the recovered gas cannot be sold, particularly from the pre-combustion system on the IGCC then who carries the cost of providing Carbon Capture? It will be the consumer, of course, and this could create difficulties for a lot of people if it results in an excessive and permanent rise in the price of electricity.

What we are now seeing is a steady stream of rejections for new coal-fired plants on grounds of costs and because they do not have the available proven technology to strip carbon dioxide out of flue gases.

Legislation particularly over the last four years has given the lie to the Green arguments that the United States doesn't care about global warming because of their refusal to sign up to Kyoto. The most encouraging sign is that the Federal Government appears to have seen through the folly of giving up nuclear power and set the conditions in Law that will support the construction of new plants and carry the technology to close the fuel cycle, and develop smaller, intrinsically safe designs. which can be applied to smaller networks and industrial power systems.

Nuclear energy has the potential to cut much of the carbon emissions from power generation and large process industries. But it requires significant public education to accept the small, intrinsically safe gas-cooled reactors with their high process temperatures in a wider range of applications.

It seems that the momentum will be carried forward under the new Government in 2009 because it has stated its determination to cut fuel imports and that may also mean encouraging hybrid car development to reduce demand for gasoline. Energy may see more emphasis on wind and Solar systems which could lead to increased production of photovoltaic cells and a wider spread of application.

So the predictions of electricity supply in 2030 are by no means firm and there could be less coal and gas, and more nuclear in a wider range of applications, particularly in the production of hydrogen as a transport fuel to supply fuel-cell powered vehicles, in order to achieve a much reduced demand for gasoline.

10
What is in the future?

Anyone reading this book within five years of its publication will almost certainly not live to see the year 2100. The youngest could by then be over 110 years old. But for however long they live the quality of life that they will experience will depend on decisions taken in the first twenty years of this century, and critical to that is having enough food to eat and a secure supply of energy.

The other issue is the reduction of biodiversity and the extinction of species through loss of habitat and of food. This is undoubtedly due to the rapid growth of the world population. More people mean a greater demand for water, for food and for energy which until now we have largely been able to provide, but for how much longer?

Not only that, but in 2009 the world is now gripped in a financial crisis which was triggered by widespread bank mismanagement in the United States and Europe, and the effects have spread throughout the global banking system. The one benefit may be that low interest rates will continue for several years, which would help the development of the electric power system

Climate change is not the extrapolation of a small increase in temperature which has grown relatively slowly in the last 150 years, but rather the result of events which enhance or destroy the local environment, such as deforestation or change of land use which increase the discharge of the greenhouse gases. Well known examples are the reduction of the surface areas of the Aral Sea, in Russia, Lake Chad, in North Africa and Lake Mead, in the United States, and the destruction of millions of hectares of rain forest in South America, Africa and Indonesia, and which can all be attributed to the demands of an increased population.

The issue at the start of the twenty-first century must be not whether it will be unbearably hot in 100 years time but that the world population, which has trebled in the last 60 years, will not have trebled in the next 60, which would bring it to more than 20 billion by the end of the century; and that there will still be enough food, water, and energy to sustain our standard of living.

What will be the factors shaping future demand for electricity and what economies in its use can be achieved? Conservation is already at work in the important areas of lighting and thermal insulation. Several countries are now marketing low energy fluorescent units to replace the traditional incandescent light bulbs, which are now being withdrawn from sale after more than a hundred years of regular use,

Energy technology is at a cross roads. The need for energy is essential to our way of life, and what matters now is that we must produce it without damaging the environment. Electricity is the one technology where we can make a significant improvement in our condition in a relatively short time, while at the same time catering for an increased energy demand from a larger population.

Per capita consumption of electricity has not significantly changed in the last thirty years but there are more people around. If the population continues to increase to 20 billion by the end of the century what would be the global demand for electricity?

If we take a median figure of 5000 kWh/year per capita, based on present European consumption, then 20 billion people would consume 100 000 Twh per year, which would have to be supplied by 12 000 GW of plant that must be available for 365 days a year. Looked at another way, that is 7500 examples of the EPR 1600 MW reactor or equivalent must be in operation by the end of the century.

Even if the average per capita consumption over the world is only 2000 kWh/year it would still mean 3000 EPR would have to be built or 32 per year brought into operation. Given that the present order book of eighteen units is unlikely to be in service much before 2020, we are already behind schedule. It would be an impossible task if it were all to be done with 5 MW offshore wind generators where 320 units would provide the same energy, assuming the wind was blowing continuously, as one of the nuclear plants.

Kyoto convinced European Governments of the arguments for renewable energy and conservation. But there is an underlying contradiction in terms. Conservation is an understandable concept: you use less energy to achieve the same end result. But to build our future energy system on renewables we require per megawatt of capacity, at least two-hundred times as much steel, and more than a hundred times as much cable for a grid connection. Is this really sustainable; is this what the Green argument is really all about, or is it just one gigantic confidence trick?

Nine years into the twenty-first century two big energy issues have emerged: the revival of nuclear power, and the cleaning up of coal. There is growing public support for nuclear energy which is not matched by that for coal. Globally coal accounts for nearly 40% of electricity generation. But against that is the requirement for clean combustion at a time when there are growing moves to curtail emissions of carbon dioxide.

The world has been through a period of political manipulation of energy supply, with the Green infiltration of political parties, notably in Germany, which stopped development of PBMR, the importance of which was that it was an intrinsically safe, small gas-cooled reactor which could work with a Thorium fuel cycle and be installed anywhere in the world as a clean emission-free source of energy. Green protest has effectively delayed this development by more than thirty years, since the prototype units in South Africa and China will not be in operation much before 2014.

Politicians hold the power to get things done; but the big problem is to build a low-carbon electricity supply system which can meet higher levels of electricity demand from an increasing population and with the same reliability as now.

The Green Movement's argument which says that more efficient use of energy, combined with improved insulation and supported by renewable energy is not practical. There is not the industrial capacity to achieve it, as shown by the low contribution it makes to existing electricity supply. These same people advocate the practice of CCS (carbon capture and sequestration) and not just for coal-fired power plants but everything burning fossil fuels. Does this include the central heating boilers in our homes?

It is easy to look at an established industrial practice and suggest that it could be applied to power plants. But CCS would have to be applied on a much larger scale; the technology for that is not proven. We don't know what it will cost to operate, and we don't know when it could be applied.

So this again is a non-starter. To use thousands of MWh of energy and millions of dollars building systems to add on to coal-fired plants so that they produce less energy from the same amount of fuel, but with only 10% of the emissions and a much higher cost per kwh for the power produced, is nonsense. If we are serious about reducing emissions, then we should not be burning coal to generate electricity at all.

The other issue is how do we move around the world? Batteries are not sufficiently developed to be an alternative propulsion system for cars. Even if this were possible and an electric car with a top speed of 150 km/h and a range of 500 km before it had to be recharged, were to be developed in the next twenty years, the power plants would have to be built to supply random charging energy to several billion vehicles.

Is an electric car which does not match the performance of present gasoline-engined cars an acceptable concept? If nothing else we have to build the power plants to charge the batteries as the market develops. If instead the power were to be generated on board, with a fuel cell, then the hydrogen fuel has to be generated by electrolysis of water, which would require more power plants to supply the energy to produce it

So the issue is not what sort of car we drive but rather that it will require electricity either to charge batteries or to produce and deliver the fuel. Electricity demand at the end of the century, regardless of the level of global population, will be substantially higher than it is now, because of the increased demand for transport. Even if we don't travel so much by car, but use high-speed trains to a far greater extent than at present, this will be a further increase in demand for electricity.

The alternative low carbon electricity supply is available with the present technology, which we have largely ignored in recent years. There is growing public acceptance of nuclear power which has been an emission-free source of electric power and heat for at least 60 years. License extensions, particularly in the United States, mean that nuclear plants brought into service after 1970 will not shut down until at least 2030. This will give more time for new plants to be built and the coal-fired capacity to be shut down.

The ideal low-carbon system would comprise nuclear power for base load, combined cycle for mid and peak load with hydro for peak and stand-by duty, and such other renewables as are available. Combined cycle already has much lower emissions than the best of the present coal-fired plants, and their flexibility of operation would mean that they need not operate for much more than 14 hours/day

This low carbon energy system can be also more economic in materials and energy use in construction because it is built up from a few large units of high output and not large numbers of small units of limited output and availability. In fact the less energy we put on certain renewables the more energy will be saved from the one-off processes of manufacture and construction of each generating plant. After all, a program to reduce energy consumption must also include greater energy economy of manufacture.

A further route of technology would be to recycle the millions of tons of plastic scrap from bags to bottles and convert it to diesel fuel. The paper industry uses more recycled paper than in the past although some virgin pulp from sustainable forests is still required for the best grades.

For low-carbon energy the technological challenges are to produce more efficient gas turbines and smaller gas-cooled nuclear reactors which will open up the industrial market particularly in operations such as coal gasification; any process in fact which must produce the required energy by burning some of the feedstock. As an electricity generator it can be applied to smaller networks and Island systems so that many more people can have access to emission free electricity supply.

Electricity supply is not the only combustion process to be cleaned up. Already some industrial emissions have been reduced as the result of deregulation and the market it created for combined heat and power. However, deregulation has not seen an increase in district heating. Some European networks were extended in the 1970's following the oil crisis of that time, but electricity supply is now focused on generation technologies to reduce emissions and improve efficiency.

The public supply of heat is unlike the other privatized energy services. The gas pipe entering the house and the electric power cable are simply the terminals of a public network to which several companies supply to others that deliver to individual consumers. District heating is a network in a small area tied to one particular power plant and consumers cannot change supplier as they can for electicity or gas.

However a wider application of district heating and cooling would concentrate emissions unless a nuclear heat source were to be used, when there would of course be no emissions. Thirty years ago district heating from nuclear power plants was being discussed in Sweden, Germany and Switzerland, but the only schemes ever built were a small one based on the Beznau nuclear station near Baden, Switzerland. A larger scheme, serving three towns in western Slovakia, nnd based on the Bohunice nuclear plant has been operating since 1997.

Houses of the future may be somewhat different. New houses will be built with double or triple glazing, cavity wall and roof insulation to reduce energy demand for space heating, and have low energy lighting units. But what will be the method of heating?

Will the default system still be a condensing gas boiler, or will the lower demand be met by a micro-CHP unit? Such a system would generate electricity which, when not required by the householder, would be sold to the public supply. Hot water for washing and space heating would be supplied by the heat recovery boiler.

Alternatively, with the development of lower cost photovoltaics, there is the potential to install small assemblies on the sun-facing roof of every new building. This could be written into building codes so that solar electric units would be installed as appropriate on all new houses.

Equally either photovoltaics or the micro-CHP system could be back-fitted on existing houses.

But before this can be done there must be proper legislation in place to define the rights of the householder and the conditions under which he can trade electricity. There is also the need for intelligent meters which can record power imported and exported. But it also requires an intelligent grid to be able to respond to thousands of consumers with just a few kWh to sell at random.

The intelligent grid is already being planned as part of the investment in refurbishment. Building new power stations on the sites of old plants is not confined to nuclear stations. Combined cycles built on the sites of old steam plants are numerous and take advantage of the existing grid connexions. In fact grid voltages have not increased since the 1970's when the existing 735, 500 and 400 kV systems were installed.

Look back fifty years, to a time when the first nuclear power stations were coming into service. Coal was losing one market after another: railways, ships, town gas, and domestic heating. Whatever was put in its place whether oil, natural gas, or electricity, was cleaner and easier to use and did not have the high level of gaseous emissions and of ash, associated with coal. It seemed that nuclear energy might even remove coal from power generation.

It didn't happen, because, the introduction of larger generating sets with higher steam conditions increased the efficiency of coal firing and gave it a competitive advantage. FGD and other measures in the 1980's added to the cost so that nuclear power even with the previous generation of reactors now had a competitive advantage which can only increase as environmental charges for carbon emission are piled on to fossil fuels.

Now with concern over climate change the simplest way to cut carbon emissions would be to remove coal from electricity generation altogether. Even with gas-fired combined cycle, there would be a significant reduction of emissions and in the short term this could bridge the gap between closure of a coal-fired plant and the completion of the nuclear plants to replace it.

Gas is the only fuel for which a practical CCS process has been developed and put into commercial operation. There are industrial combined heat and power schemes mainly in the Asian fertiliser industry where there is a commercial value for the recovered gas. This could be the only application of CCS other than possibly slip-stream applications on large steam plants near to oilfields: where there is a commercial market for the carbon dioxide tor enhanced oil recovery.

This is all that we can do to have an immediate effect because we still have to cope with the problems of oil as a fuel for transport. There are many professionals who depend on large diesel engines to drive trucks or tractors or fishing boats, so that the price of oil, which in 2008 shot up to over £140/bbl and has fallen back to to around $70/bbl, is in all countries a critical factor in the cost of food production and general transport which cannot be compromised.

Electric vehicles have been around a long time for local delivery work, and use rechargable batteries as the power source, which limits their range and speed with the regular starting and stopping. An electric car would not have the same pattern of operation and would be a somewhat lighter vehicle. Some hybrid vehicles have gasoline engines charging lithium-iron batteries for the electric drive motors. But a car which is not compatible in performance with current vehicles, particularly in its range, is unacceptable.

In California hydrogen filling stations are now available and can supply safely liquid hydrogen to power a fuel cell which drives the electric motors. Cars so equipped have a range of at least 500 km and acceleration to match a gasoline-engined vehicle. To extend this market will require power plants to produce the hydrogen and the combustion product from the cars will be water vapour. So that while, today it is marginally warmer in a city than in the surrounding countryside, in future it may be more humid as well as the hydrogen-powered vehicle population increases.

To produce the hydrogen will require energy and the logical method is by electrolysis of water. This can be done by a nuclear plant or the intermittent operation of wind farms could cause them to be detached from the grid and dedicated to hydrogen production. The hydrogen distribution system must be based on present filling stations, as is starting to happen in the United States.

Bio fuels have been with us a long time, particularly in Brazil, where ethanol made from bagasse dates from the first oil crisis of the 1970's. Biomass power plants burning wood chips and sawdust, or crop residues such as bagasse and rice husks are using waste plant material, and are generally built close to their source

But how much land, if any, can we give over to grow biomass for fuel production in the face of a growing global population. It's all very well to say that today only 3% of the world's arable land has been given over to bio fuels, but how much will be used by 2050 when there might be half as many more people on the earth as there are now.

The basic problem is that Green influence has been moving us in false directions and we are beginning to see where these are leading us. Coal may be the most abundant fossil fuel but it is the least efficient generator of electricity. It has done much environmental damage with acid rain and smog, and the effects on public health.

Yet to strip out carbon dioxide from the flue gases, using a process which is only now at the pilot-plant stage, creates more problems than it solves and to do that on every fossil-fired power plant and industrial boiler "to save the planet" is not practical.

A Green Electricity Supply System which does not damage the environment can be built on the existing technology. Whatever is the future demand for electricity it will surely include hydrogen production to supply a growing global vehicle population.

Nuclear construction is expected to pick up after 2010. Five years later, new, smaller, gas-cooled reactors will have the potential to extend application of nuclear energy to industrial processes, district heating, and hydrogen fuel production.

In two hundred years time will historians look back on this time as a period in which our political leaders had such a poor grasp of technology and some were so fearful of it that they were more concerned to placate public opinion rather than to lead it. We have had the technologies to create a low carbon energy system, for fifty years but lacked the will, or should one say the courage, to implement it.

Finland's government decided almost as soon as the ink was dry on the Kyoto agreement that the only way to meet their emission targets was to build another nuclear power plant. Why did none of the other Kyoto signatories follow suit? In fact, they have all gone down the renewables route and some look on nuclear as an option of last resort.

Yet if we look to the Far East and the ongoing nuclear energy programmes of the major powers there, it is evident their aim is to have a green energy system in the future, which is capable of meeting the projected electricity demand.

The rest of the world has to wake up to the fact that we cannot continue to burn fossil fuels any more than we have to. We should be working to set up the truly green energy system with the minimum emissions and the maximum efficiency. It must be able to meet the needs of society with an adequate reserve margin for security, and minimum impact on the environment. This is what a green energy system is really all about.

Index

Alstom, 49, 50, 51, 66, 78, 90,
119, 145, 147, 152, 206, 227
American Electric Power (AEP),
USA, 47, 51, 52, 70
American National Power
(ANP), USA, 140, 141, 147, 206
Ansaldo Energy, Italy, 163
Areva, France, 84, 95, 103, 108-111,
112, 114, 116, 122, 124-126,
208, 209, 210, 212
Atomic Energy of Canada Ltd
(AECL), 110, 112, 114, 122, 124
Bechtel Corporation, USA, 70, 72,
205
Benson, Mark, 62, 147
Benson boilers, 157-60, 162, 163,
165-168, 170, 172-174, 180
Biomass co-firing 78-79
Biomass power, 175-176
Blackburn Meadows, UK, 176
Bonneville Power Administration,
USA, 204
British combined cycle plan, 151
British Energy, 109, 123, 124, 125
British Nuclear Sites, 125
Bruce Power, Canada, 123
Bureau of Reclamation (US),
202, 204
Carbon Capture and Storage
(CCS), 43-58, 71, 77, 88, 205,
212, 214, 220, 223, 224,
CCS effects on performance, 48
Chilled ammonia process, 47-51
Centrica plc, 57, 123, 125, 137, 151,
191
China Nuclear Power Engineering
(CNPE), China, 112, 113
Chinese nuclear plans, 112
Coal-fired steam plants
Castle Peak, Hong Kong, 61
Drax, UK, 79
Hamm, Germany, 66

Mountaineer, USA, 51-52
Oologah, USA, 52
Pleasant Prairie, USA, 47, 49-52,
212
Schwarze Pumpe, Germany, 64
Yuhuan, China, 65
Coal Specifications, 71
Cockerill Mechanical Industries
(CMI), Belgium, 133
Combined Cycle
Angleur, Belgium 136
Baglan Bay, UK, 147, 148
Bang Pakong, Thailand, 137, 140
Blackstone, USA, 141
Cottam, UK, 144
Damhead Creek, UK, 151
Dighton, USA, 142
Higashi Niigata, Japan, 135
Mongstadt, Norway, 43
Paka, Malaysia, 136
Irsching, Germany, 148, 149
Roosecote, UK, 137
Tung Hsiao, Taiwan, 136
Combined Heat and Power,
9, 10, 38, 39, 44, 47, 57, 65, 68,
82, 91, 107, 109, 128, 131, 134,
136-138, 142, 151-152, 160-161,
168, 171, 177, 199, 200, 211-212,
221, 223
Combined Operating Licence, COL,
107, 207, 209
Constellation Energy, USA, 121
Department of Energy (DoE), USA,
11, 38, 53, 54, 80, 214
Dong Energy, Denmark, 45, 191
Doosan Heavy Industries, Korea,
111, 113, 114
DOWNViND programme, 179
Edf Energy, UK, 79, 109, 123, 144,
151, 169
Eisenhower, Dwight D, President
Atoms for Peace policy, 89-90

Electricité de France, 9, 79, 103, 109, 109, 121, 123, 124, 125, 153, 188
Electric Power Research Institute (EPRI), USA, 11, 49, 52
EnCana Corporation, Canada, 44
Entergy, 121, 208, 209
E.ON, Germany, 148, 207
E.ON, UK, 125, 151, 169, 175-176, 189
European Marine Energy Centre (EMEC), UK, 166-167, 188-189
Exelon Corporation, 120, 209
Fermi, Enrico, 83, 85, 90
Flue gas desulphurization (FGD), 4, 5, 12, 47-48, 53, 57, 60, 69, 70, 74, 87, 89, 90, 92-93, 151, 226
Futuregen, USA, 53, 54
General Electric (GE), USA, 67, 70, 72, 84, 90, 104, 108, 110, 121, 123, 136, 139, 146, 148, 199, 206, 210, 230
Geothermal sites
  Geysers, USA, 163
  Habanero, Australia, 164
  Jolioka, Australia, 164
  Lardarello, Italy, 163
  Soultz sur Foret, France, 165
  Wairakei, New Zealand, 163
Global Nuclear Energy Partnership (GNEP), 106, 107, 161, 210, 211,
HSBC, UK, 172, 173
Hydro power plants
  Aswan High Dam, Egypt, 186
  Blantyre, UK, 184
  Dinorwig , UK, 185
  Glendoe, UK, 186
  Grand Coulee, USA, 202
  Grand Inga, Congo, 186
  Hoover Dam, USA, 203
  Inga, Congo, 186
  Romney Weir, UK, 183
  Three Gorges, China, 113, 182, 186
IEA Greenhouse Gas Research and Development, UK, 44, 45

IGCC projects
  Iwaki, Japan, 69
  Kellerman Lünen, Germany, 67, 69
  Schemes in Operation, 69
  Vresova , Czech Republic, 69, 78
  Wabash River, USA, 68, 69
International Atomic Energy Agency (IAEA), 120
Kansai Electric, Japan, 54
KM CDR process, 54-55
KS-1 Absorber, 54
Korea Midland Power, 189
Korean nuclear plans, 114
Krechba Gas Field, Algeria, 43
Kyoto Conference, 10, 11, 17, 22, 24, 31, 32, 36, 38, 42, 52, 105, 170, 195, 211, 215, 219, 225
Large Combustion Plant Directive (LCPD), 11, 37, 64, 77, 80, 82,
Marine Current Turbines, UK, 166, 168, 187, 189
Marine Tidal Energy, Canada, 189
Mitsubishi Heavy Industries, Japan, 54, 69, 70, 103, 108, 110, 112, 113. 116, 135, 142, 158, 208
Nuclear Decommissioning Authority (NDA), UK, 124, 125
Nuclear power plants
  Balakovo, Russia, 119
  Bellefonte, USA, 121, 209
  Beloyarsk, Russia, 89, 100, 119, 129
  Biblis, Germany, 85, 109
  Braidwood, USA, 121
  Bruce, Canada, 123
  Calder Hall, UK, 2, 8, 90, 91
  Chernobyl, Ukraine, 20, 30, 99, 100, 102, 105, 160
  Civaux, France, 103, 125
  Darlington, Canada, 123
  Daya Bay, China, 111, 113
  Diablo Canyon, 96, 211
  Dresden, USA, 210
  Fessenheim, France, 125

Flamanville, France, 37, 110, 125, 126
Gosgen Daniken, Switzerland, 91
Grand Gulf, USA, 121, 209
Gravelines, France, 103, 125
Haiyang, China, 111, 112
Hinkley Point, UK, 124, 125
Hongyanhe, China, 112, 113
Ignalina, Lithuania, 101, 103, 129
Kalinin, Russia, 118, 119
Kashiwazaki Kariwa, Japan, 104, 118, 207
Koeburg, South Africa, 107, 126, 128, 157
Kola, Russia, 119
Kori, Korea, 113-115
Ling Ao, China, 84, 111, 112
Loviisa, Finland, 102, 103
Lungmen, Taiwan, 84
Novovoronezh, Russia, 119
Okiluoto, Finland, 33, 110, 111, 140
Pickering, Canada, 138
Quinshan, China, 112, 128
Sanmen, China, 110, 112
Seversk, Russia, 119
Shin Kori, Korea, 114. 115
Shin Ulsin, Korea, 114, 115
Shin Wolsong, Korea, 114 115
Shippingport, USA, 2, 90, 91, 204
Sizewell, UK, 21, 124, 125
Smolensk, Russia, 119
Sosnovy Bor, Russia, 119
Susquehanna, USA, 121
Three Mile Island, USA, 97, 99, 160
Tianwan, China, 112, 120
Temelin, Czech Republic, 95, 116
Ulchin , Korea, 114
Volgadonsk, Russia, 19
Watts Bar, USA, 83, 105, 110, 201, 206, 211
Wolsong, Korea, 114
Yonggwang, Korea, 108, 114, 115
Nuclear Reactor types

ABWR, 110, 117, 118, 209
ACR, 1000, 89. 123
AGR, 125, 130
AP 600, 110, 161
AP 1000, 108, 110, 112, 123, 207, 209
APR 1000, 137, 145
APR 1400, 114, 115
BWR, 103, 104, 109, 121, 230, 231, 233
CANDU, 99, 101, 105. 108, 109, 119, 121,126, 129, 114, 138, 139, 180
CNP 600, 126
CPR 1000, 126, 128
EPR, 95, 103, 108, 110, 112 121, 123, 124, 125, 129, 208, 209, 212, 218
ESBWR, 108, 110, 121, 123, 207, 209
Fast Breeder Reactor, 101, 114, 134
Magnox, 90, 96, 99, 104, 105, 131, 139, 140, 146
PBMR, 99, 121, 135, 143,144,145, 146, 176-180, 232, 242
IRIS, 121, 122, 136, 179-180, 233
OPR 1000, 114, 115
Pressurized Water Reactor (PWR), 2, 42, 96, 101, 103, 104, 116, 117, 118, 119, 120, 121, 124, 126, 127, 128, 129, 114, 132, 133., 134, 136,137, 139, 142, 144, 175, 179, 180, 223, 225, 227, 228, 230, 233, 234, 235
RBMK, 99, 114, 117, 134, 145
System 80 (KSNP), 130
Nuclear Regulatory Commission, USA, 230
Nuclear ships
  Lenin, Russia, 92. 93
  Otto Hahn, Germany, 106
  Savannah, USA, 106
  Sevmorput, Russia, 93
Pennsylvania Power & Light (PPL),

USA, 121, 123, 199, 209
Progress Energy, USA, 93, 229
Pumped storage, 184-185
RE Power Systems, Germany, 179
Repowering Steam Plant
    Drogenbos, Belgium. 133
    Rhinehafen, Germany, 140
    Senoko, Singapore, 143
Rosatom, Russia, 93, 112, 118
Rotech Tidal Turbine, 188-189
RWE, Germany, 66, 109
RWE Npower, UK, 124, 125, 151,
    183, 184
Sonatrach, Algeria, 42
Scottish and Southern Energy, UK,
    179, 186
Shaw Group, USA, 81, 122, 209
Siemens, 78, 64-66, 70. 72, 73, 103
    104, 108, 135, 136, 139, 140, 142,
    147-149, 155, 180, 206
Sleipner Gas Field, Norway, 43
Snøhvit Gas Field, Norway, 43
SPE, Belgium, 125
Statoil, Norway, 42, 43, 199
Stevens Croft, UK, 175, 176
Tennessee Valley Authority (TVA),
    USA, 121, 206, 207, 211
Teollisuuden Voima Oy (TVO),
    Finland, 33, 111
Tidal power plants
    Annapolis, Canada, 186, 189
    La Rance, France, 187, 188
    Severn Barrage, UK, 208
    St David's Head, 168, 189
    Skerries, 168, 187, 190
    Strangford Lough, UK, 207-208
Toshiba Corporation, Japan, 105, 108
Troll Gas Field, Norway, 44
Unistar Corporation, USA, 121, 209
US electricity production forecast,
    215
US Generating capacity forecast, 213
VAX steam turbine, 141, 142
Wartsila Oy, Finland, 93
We Energy, 49, 51, 212

Westinghouse, USA, 81, 90, 102,
    103, 107, 108, 110, 112, 113, 114,
    116, 120, 122, 123, 128, 133, 142,
    146, 158, 160, 161, 204, 206, 207
    208, 211, 230
Weyburn Oil Field, Canada,
    44, 56, 65, 215
Wind farm
    Burbo Bank , UK, 198
    London Array, UK, 179, 191-193
    Talisman , UK, 199